J.J. Thomson and the Discovery of the Electron

Thomson in the 1920s

Cathode Rays.

J.J. Thomson. M.A. F.R.S. Cavendish Professor of Experimental Physics Cambridge.

The experiments* [foot-note] discussed in this paper were undertaken ~~were undertaken~~ in the hope of gaining some information as to the nature of the Cathode Rays. The most diverse opinions are held as to these rays: according to the almost unanimous opinion of German Physicists they are due to some process in the ether to which, inasmuch as in a uniform magnetic field their course is circular and not rectilinear, no phenomenon hitherto observed is analogous: another view of these rays is that so far from being wholly etherial they are in fact wholly material, & that they mark the paths of particles of matter charged with negative electricity. It would seem at first sight that it ought not to be difficult to discriminate between views so different, yet experience shows that this is not the case as amongst the physicists who have most deeply studied the subject can be found supporters of either theory.

The electrified particle theory has for purposes of research a great advantage over the etherial theory since it is definite & its consequences can be predicted; with the etherial theory it is impossible to predict what will happen under any given circumstances as on this theory we are dealing with hitherto unobserved phenomena in the Ether, of whose laws we are ignorant.

The following experiments were made to test some of the consequences of the electrified particle theory.

Charge carried by the Cathode Rays.

If these rays are negatively electrified particles then ~~if we catch them in~~ when they enter an enclosure they ought to carry into it a charge of negative electricity. This has been proved to be the case by Perrin who placed in front of a plane cathode two coaxial metallic cylinders which were insulated from each other: the outer of these cylinders was connected with the earth, the inner with a

*Some of these experiments have already been described in a paper read before the Cambridge Philosophical Society [Proceedings. vol IX. 1897] and in a Friday Evening discourse at the Royal Institution [Electrician May 21. 1897]

Phil Mag Vol 44, 1897, p. 293 ff

ge of cathode ray manuscript in Thomson's own hand

J.J. Thomson and the Discovery of the Electron

E.A. DAVIS and I.J. FALCONER

UK Taylor & Francis Ltd., 1 Gunpowder Square, London EC4A 3DE
USA Taylor & Francis Inc., 1900 Frost Road, Suite 101, Bristol, PA 19007

British Library Cataloguing in Publication Data

A catalogue record for this book is available from the British Library
ISBN 0-7484-0696-4 cased
0-7484-0720-0 paper

Library of Congress Cataloging Publication Data are available

Cover design by Youngs Design in Production

Typeset in Times 10/12pt by Keyword Publishing Services Ltd.

Printed in Great Britain by T. J. International Ltd, Cornwall

To Christine and Kenneth

Contents

Foreword

David Thomson

The Master's Lodge at Trinity College in Cambridge faces onto the Great Court of grand and pleasing proportions and is backed by its walled garden leading down to the River Cam. It has been the serene home for more than three hundred years of great academic figures. Sir Joseph Thomson, my grandfather and the last Master to hold the position for life, died there in August 1940 at the age of 83. He and his wife Rose had lived in the Lodge, together with their daughter Joan, for 22 years, presiding over the grandest of Cambridge Colleges with a simple wisdom, tact and humour that complemented his achievements as a physicist of 'super-eminence', as Lloyd George once described him.

1940 was a bleak moment for the nation and scarcely an appropriate time to assess the place in science of 'J.J.' Thomson, the initials by which he was invariably known rather than the title Sir Joseph. The main period of his creativity lay roughly between the years 1890 and 1910; the high point, which gave him fame, was his work with cathode rays leading in 1897 to his theory about 'corpuscles smaller than atoms', later to be called electrons, and subsequent confirmation of these theoretical ideas with proof of the particles' existence, which many had doubted, by a series of brilliant experiments.

The setting for these discoveries was the Cavendish Laboratory of which J.J. was professor for the long span of 35 years. Apart from his own groundbreaking work which took the first steps into the subatomic physics of the twentieth century, he directed, recruited and inspired a succession of brilliant young men round him at the Cavendish who were to share this work and later to create whole new fields of science. It is probable that there has never been such a successful and sustained record of new scientific discovery in one laboratory across such a broad field as that which happened under J.J.'s leadership of the Cavendish laboratory from 1884 to 1919. It can claim to be the first real

School of Physics in the modern sense, and in his time it was generally regarded as the leading experimental laboratory of physics in the scientific world.

In 1997 we are celebrating the centenary of J.J.'s first exposition of his thesis that a cathode ray particle is more than 1000 times lighter than the lightest chemical atom and a universal constituent of matter. It is also an opportunity, now long overdue, to re-assess J.J.'s own reputation and place in the history of science, judged against the developments throughout the twentieth century both in pure and applied science which built on the foundations of his work and that of the group round him at the Cavendish.

At J.J.'s death his great period of work had ceased a whole generation earlier. He was certainly not a man with only one discovery to his name, as his subsequent work on positive rays leading to the discovery with Aston of non-radioactive isotopes, and the constant ferment of his intellectual energy amply proved. However, by the time the First World War started in 1914, when he was nearing sixty, his main work had been achieved.

The effect of that war on the laboratory practically brought original research to an end. The younger scientists, including J.J.'s son, soon found themselves in France, fighting at the Front. J.J. himself was drawn into War Committee work and also became President of the Royal Society. Shortly before the end of the war Montagu Butler, then Master of Trinity, died and J.J. was appointed to succeed. He did not therefore take up an active research role when peace time conditions returned and he retired officially from the Cavendish in 1919, though retaining an honorary professorship and space for his own research. His most famous pupil and collaborator, Ernest Rutherford, succeeded him and J.J. became increasingly a senior spokesman of Science. He did however continue to produce new work, much of it based round universal theories of the ether, harking back to some of his earliest concepts.

In 1940 then, the formal honours were paid at his death. He was a national figure. In the popular mind 'the man who split the atom', he was the senior member of the Order of Merit, the epitome of Cambridge science and something of a legend, in Cambridge at least, as a much admired, if sometimes absent-minded, genius.

His ashes were buried in Westminster Abbey and the stone above them lies in the shadow of Isaac Newton's memorial and next to that of Rutherford, who had predeceased him, sadly early, three years before.

Trinity did not lack for great names to succeed him. After science, it was the turn of history. The appointment at Trinity is a royal one on the Prime Minister's recommendation, and the choice most aptly was George Trevelyan. To quote Trevelyan, 'a man's reputation with the world at large is usually at its lowest exactly a hundred years after his birth, especially in the realm of letters'. It was to be true of Trevelyan himself, and in the realm of science equally true of his predecessor at Trinity. In both cases the tide is turning.

J.J.'s background was Manchester; liberal, mercantile Manchester, where his family had lived for three generations. His great-grandfather came from Scotland – hence the Scottish spelling of Thomson – and started a small business selling antiquarian books, occasionally acting as a publisher too. It must have prospered sufficiently to give a reasonable living and a house with servants in the middle-class suburb of Cheetham. Socially this would have put the family in the merchant middle class, somewhat below the doctors, solicitors and clergy of the professional classes.

J.J.'s father, also Joseph, died aged 38 when the boy was only 16, and the only other child, Frederick, four years younger. Though not much seems to be remembered of the father, it is recalled that he was proud to be acquainted with some of the scientific and literary people of Manchester, many of whom would probably have known his bookshop. Indeed it was in this way that he came to hear of Owens College and its growing reputation for good teaching. The father seems to have decided that his son's aptitude for mathematics suggested that engineering was a more suitable choice than bookselling, perhaps hoping that the younger son would follow in the family line. As the engineering apprenticeship to a Manchester firm of engine-makers could not be started until 16, and J.J. had precociously outgrown his school by 14, Owens College would fill the gap.

J.J. in later days regarded this as the lucky chance that shaped his life. When his father died unexpectedly after two years at Owens, there was no money to pay for the apprenticeship; the business was sold, his mother, Emma, moved to a smaller terraced house nearer the College and found enough money to keep J.J. there.

Manchester had of course a great tradition of science. Its Literary and Philosophical Society founded in 1781 was older than the corresponding Royal Institution in London. John Dalton, the great chemist and author of the atomic theory, had his laboratory in the Society and was its servant as Secretary and then President from 1800 to 1844. He was succeeded by James Joule who, at the early age of 22, had discovered the law named after him and, at 25, had published the paper which led to the acceptance of the principle of the Conservation of Energy. J.J. was to write in his autobiographical *Recollections and Reflections* that this discovery was 'a striking instance of how great generalisations may be reached by patient work', a phrase that might be used of his own work on the electron. When J.J. was a boy he had been introduced to Joule by his father who had said afterwards 'some day you will be proud to say you have met that gentleman'.

It was this sort of influence and background that shaped the schoolboy's thoughts to science. He was a shy boy but determined and 'knew which way he was pointing'. It was at about this time that when asked by a cousin what he wanted to do when he grew up, he replied that he intended to go in for *original research* – a rare ambition in 1870 – and his cousin remembered 'my brother-in-law tapped him on his head and said "don't be such a little prig, Joe".'

Owens College, called after the merchant who left his money to found it in 1851, had a shaky start but twenty years later when J.J. arrived it had started to prosper. There was probably no other college outside Oxford and Cambridge that could match it in England, though Glasgow was perhaps its equal in Scotland. Later, in his *Recollections*, J.J. was to write 'though Owens College was badly housed, no university in the country had a more brilliant staff of Professors'. The range of lectures in engineering, mathematics and 'natural philosophy', as physics was then called, was impressive. In effect he had been through university courses in three subjects while still a schoolboy, and thrived on it.

He needed to proceed by scholarships; there was no money otherwise to go forward academically, and his virtuosity had made it clear to his teachers that Cambridge, with its great reputation for mathematics, was the place to aim for. The route had to be mathematics; there were no entry scholarships for engineering or physics. He succeeded at his second try at Trinity, the College with the highest reputation and the hardest to get into by scholarship. There were by then a number of reasons pointing to Trinity. The first was his own self-confidence in his ability; he knew he was in the top division and did not hesitate to test himself. The second was that a number of his teachers had been Trinity men, particularly Thomas Barker, the professor of Mathematics, who had also been Senior Wrangler – that sonorous intellectual title of Victorian academia. And a young member of staff, John Poynting, arriving at Owens shortly before J.J. left and one of his closest friends in later years, was also from Trinity. Not long before, John Hopkinson, a future Professor of Engineering at Cambridge, had left Owens for the same destination and won distinction as Senior Wrangler. It all seemed to point in the same direction.

Perhaps a final reason for a young man with his eyes on *original research*, and who had published his first short paper in the *Proceedings of the Royal Society* when aged 19, before leaving Owens, was that Trinity had famously been the home of Sir Isaac Newton and was currently the College of James Clerk Maxwell, the first Cavendish Professor of Experimental Physics, whose famous *Treatise on Electricity and Magnetism* was a starting point for much of J.J.'s early thoughts about research.

J.J.'s personal life divides quite simply; his family background and education in Manchester, and his subsequent life in Cambridge. In his long life he never lived anywhere else but those two places. To round off the Manchester period, it would be appropriate to say something of the family he was about to leave behind. His relations with his mother had always been close. She was a sweet-natured person, proud of her son without, I suspect, understanding anything at all of his work. He spent part of his summers with her regularly, after his marriage as well, up to her death in 1901, either in Manchester or at some seaside resort. His younger brother, Fred, a bachelor all his life, was from my father's account a man of selfless decency and kindness. He seems to have accepted his brother's success with admiration, and felt no rancour that he had not had the education which, with such effort, had been afforded his elder brother. Fred worked for a firm of calico merchants in Manchester, Claflin & Co., but was never a partner in it. His real love, and success, was in his voluntary work organising boys' clubs – particularly Oldham Lads Club – and taking them on summer camps. He retired early from ill health and lived the last three years of his life in Cambridge in lodgings in Newnham, an easy walk from Joe and Rose.

In later life J.J. played up the Manchester link though he rarely revisited it. He spoke with a noticeable Mancunian burr to his speech. His robust sense of humour, stories and jokes rather than intellectual wit, was also from that background; and so too his liking for the regional dishes, Lancashire hot pot and Manchester pudding, which I can recall as a boy were still favourites of his in his eighties.

At Cambridge, his personal life can be divided into three periods. The first covers the student and unmarried early years (1876–89), as undergraduate, College fellow, lecturer and young professor, living in rooms in Trinity (except for the first years when he was in College lodgings in the town) up to his marriage when he was 34. The second period spans the years of his greatest creativity as Cavendish Professor. He lived the early years of married life in Scroope Terrace in the town, where my father was born. In 1899 the family moved to a large Victorian house in West Road called Holmleigh, with more than an acre of garden near the Backs on the other side of the River Cam from the colleges. The third period came with his appointment as Master in 1918 when he moved back to Trinity into the Master's Lodge for the rest of his life. All of these homes were within ten minutes walk of each other, and of the Cavendish Laboratory, which was then in Free School Lane in the old centre of the town. It was therefore a small circle geographically that encompassed his domestic life for 65 years. It suited him perfectly. He liked Cambridge as a place. The air, he said, seemed to favour him; he was never ill, not once in that time could he recall a working day lost by ill health.

The early years of J.J.'s Cambridge life up to his marriage were remarkable for the extraordinary speed with which he established himself, moving from one goal to the next. Eight years after coming up to read mathematics he was elected Cavendish Professor of Experimental Physics in succession to Maxwell, the first holder, and then Lord Rayleigh, who held the post for only five years, two great names in British physics. It was an achievement which astonished his contemporaries and continues to do so. How was it done?

Mathematics was then (at Cambridge at least) the usual approach to the study of physics, and the best students endured the extraordinary rigours of working for the Mathematical Tripos, if you aspired to be a Wrangler. With the help of the famous coach Edward Routh, J.J. emerged as Second Wrangler close behind Joseph Larmor, who was also to spend his life at Cambridge at St John's College in a career of great distinction. The two men became lifelong friends, though there is little correspondence between them to show it, mainly because they lived so close and saw each other often.

It was in J.J.'s last year as an undergraduate that Maxwell died, aged 48, after some years of ill health and periods of absence from the Cavendish. It was a matter of great regret to J.J. that he never actually met his predecessor whose work he so much admired. He did, however, acquire Maxwell's armchair, and used it for the rest of his life in his study at home; he did much of his mathematics sitting in that very chair.

After graduating, J.J. stayed on at Trinity to try for a Prize Fellowship. This was the point when he had to decide whether to be a pure mathematician or a physicist. He was in little doubt which way to go. The taste of physics at Owens College determined his choice for his fellowship thesis on the transference of kinetic energy which followed on from Professor Balfour Stewart's lectures on energy at Manchester. Three years were allowed to submit the work for a fellowship, but with great concentration and hard work, J.J. completed his thesis within a year, and with his usual self-confidence submitted it against the judgement of his tutor – and succeeded. In the following year, 1882, he won the Adams Prize on the Motion of Vortex Rings and became an assistant lecturer in the College; in 1883 he became a University lecturer. These posts secured him an income and a degree of financial security for the first time. He was now teaching in Trinity and beginning experimental work in the Cavendish: the twin poles of his life at Cambridge were falling into place.

J.J. published a remarkable number of important papers in those three years between his fellowship in 1881 and the key year of his early academic career, 1884. These papers were both theoretical and to a lesser extent experimental, including – in 1883 – the first paper on the subject which he was to pursue all his life, 'On a Theory of Electric Discharge in Gases' – a new area of research at the Cavendish and the one which was to lead to the discovery of the electron.

These early papers made his reputation in the small world – as it then was – of professional mathematicians and physicists. So much so that in the first half of 1884, unusually soon in his career, he was elected a Fellow of the Royal Society. Without that accolade it is doubtful whether a few months later, when Lord Rayleigh retired from the Cavendish, he would have been appointed to succeed. As it was, the appointment took everyone by surprise. The fact that he should have put himself forward for the post at all was further proof of his sturdy self-assurance.

There were nine electors for the post and at least four of them, in the previous three years, were directly involved in adjudicating J.J.'s work for the Trinity Fellowship, the Adams Prize or for the fellowship of the Royal Society. In particular, the external elector was Sir William Thomson (later Lord Kelvin) from Glasgow, no relation to J.J., who could himself have had the post on all three occasions if he had so wished. He was, however, such a grand figure in Glasgow with his electrical engineering business as well as his academic position that it was scarcely to be expected that he would wish to return to Cambridge. He had a commanding reputation and equally formidable manner of speech and it was probably his support for J.J. that swayed the decision in the latter's favour. The electors took a far-sighted gamble and backed the original thinker of great promise rather than other candidates with more proven experience of teaching and experimentation, two of whom had actually taught J.J. in Manchester.

The appointment made J.J. a natural leader of science in the wake of Maxwell and Rayleigh. The three men could scarcely have been more different:

> Clerk Maxwell brought the genius of the Scottish Celtic lairds and Rayleigh that of the English landed aristocracy to bear on the creation of the Cavendish Laboratory. J.J. Thomson added to this the potent spirit of the nineteenth-century British middle class. He had great scientific gifts but his shrewdness, industry, common sense and membership of that class of men who dominated the Victorian period was also very important. He was better equipped than Maxwell and Rayleigh for developing the Cavendish Laboratory after it had been started and consolidated. [1]

It took some time for J.J. to get established in the post, find lines of research and, scarcely less important, the money to enlarge and equip the laboratory. In the 1880s physics generally was passing through a dead period: there were no grand themes, but a lot of tidying up and improvement of measurements as well as a gradual improvement in the methods of teaching and examining. The number of students at the Cavendish was small, and the organisation of the laboratory informal. The professor conducted personally the administration, including it seems the book-keeping. J.J. wrote his correspondence, standing up at a desk with a sloping top, in a neat and legible handwriting with the minimum of punctuation. It was the only thing about him that was neat in fact, since he was notably the opposite in organising his papers, his timetable and his dress.

There were few formal entry requirements to be a student at the Cavendish. In the case of Rose Paget, who came as a student a year or two after J.J. became professor, there was no formal qualification at all, only an intellectual fascination with science which had been fostered by her father, and a determined will. She and J.J. married in 1890; she was the only woman in his life and centre of his affection for fifty years in a marriage of much warmth and mutual support.

Rose and her twin sister Violet were born in Cambridge in 1860 in the middle of a large family of ten children, of whom seven survived childhood. Rose and Violet – a coupling of charming mid-Victorian whimsy – remained quite devoted to each other all their lives, though the two were almost opposites in character. Rose was intellectual, precise and calmly organised, Vie scarcely educated, as she liked to say, extrovert and a passionate gossip. A good deal of correspondence between them still exists; only to look at the handwriting tells it all. Rose wrote in a fine regular script, Vie in a hasty hand that ran out of paper.

Their father, Sir George Paget, was by the 1880s something of a grand old man in Cambridge, where he had lived continuously since coming up to Caius College in the 1830s. He was the Regius Professor of Physic, that is to say medicine, a fellow of the Royal Society and, at the time Rose started at the Cavendish, had just been knighted (KCB) – a rare honour then in academic Cambridge. Originally a mathematician at Caius and a Wrangler, he switched to medicine because it offered him a tied fellowship at Caius, and had nearly been elected Master there, but was thought – unlike his future son-in-law at the Cavendish – to be too young for the post. This was personally a blow to him at the time, not least because as Master he would have been able to marry. The much older man who got the post – at that time all masterships were for life – hung on for an unconscionable time and George Paget resigned his fellowship, only months before the Master died, to marry Clara Fardell (the daughter of a clergyman in the Isle of Ely) and start a family.

Had George Paget become Master of Caius, it is unlikely he would ever have achieved his life's work which was to transform the teaching of medicine at Cambridge. He had a great reputation for getting things done in the University by a combination of common sense and a detailed knowledge of how to steer a course through the complex web of University Committees on many of which he served. At the same time he was the leading general medical practitioner in the town of Cambridge, and the surrounding fen countryside, which helped to provide the means, before he became Regius Professor, to bring up his large family in St Peter's Terrace next to the Fitzwilliam Museum and opposite Addenbrooke's Hospital, which was the centre of his activity. His better-known younger brother, Sir James Paget, was at this time the inspirational force in medical teaching in London, at St Bartholomew's Hospital, and surgeon to the Queen.

It seems surprising that the daughters of an academic family in Cambridge could have had such a casual education in the 1860s and 70s. My father wrote about his mother: 'She was almost devoid of formal education except for that provided by a few years of a German governess, and the lessons of a teacher of French named M. Bouquel who lived nearly opposite and whom I remember her visiting each Christmas with a small present.'

Later, Rose and Violet were to spend a few months in Dresden, following their elder sister Maud, to improve their German. Rose's education in mathematics and science came largely from her father, the two were very close, to the point (my father told me) of straining her relationship with her mother. Though as a boy I lived for a while with my grandmother, and heard all her remembered stories of St Peter's Terrace, I do not recall that she ever mentioned her mother, though in her eighties her feelings for her father were still palpable. To quote my father again on his mother before her marriage:

> My mother had two outstanding passions in life, devotion to her Father (I cannot refrain from the capital letter she always used for him) and an adoration of science, especially physics. She was a genuinely religious woman but I doubt if religion meant as much to her emotionally as these. Somehow she managed to learn a surprising amount of mathematics, though I think she never really mastered the Calculus. I believe she passed some of the 'Cambridge Local Board'

examinations but to what standard I don't know. Certainly her knowledge of physics would not admit her nowadays to do research in the Cavendish. To what extent her undoubted good looks may have influenced my father in admitting her I can't say, but they had only met once or twice before.

The courtship of professor and student can be followed from the small packet of letters which Rose kept all her life. It covered a period of about two years; a sample shows the progress from formality to passion.

Cavendish Laboratory
March 16th (1888)

Dear Miss Paget
I think you might take the Demonstrations on Light next term, these will require a knowledge of the Undulatory Theory of Light and if you have not studied that subject it would be advisable to do so before taking the Demonstrations. You will find all you require in Mr Glazebrook's book on *Physical Optics* in Longmans Series.
Yours very truly
J.J. Thomson

Trinity College
August 11th 1889

Dear Miss Paget
I enclose the solutions of the problems you sent and hope that you will soon send some more. I must thank you for the very large piece of bride-cake you sent. It is so large that it is not nearly done yet. I saw Miss Skjort today, she seemed in good spirits and most enthusiastic about the merits of Ibsen.
Believe me
Yours very truly
J.J. Thomson

Cavendish Laboratory
October 2nd 1889

My dear Miss Paget
I shall be very glad to talk with you about the work for next term if you are likely to be at home any afternoon except tomorrow. I will call at St Peter's Terrace for the purpose.
Yours very truly
J J. Thomson

Undated draft letter from Miss Paget

My dearest Friend
I have been thinking unceasingly on all that happened this morning and the perplexity becomes more distressing the more I think of it. I know that my feelings betrayed me this morning but I could not help it and I was not prepared, the surprise was so sudden. *Whatever* happens I know only this that I *cannot* leave Papa so that if there is to be any future for us together it

can only be with waiting. I ought to have explained it all to you so much more carefully today instead of thinking only of the present moment and forgetting myself altogether. Please forgive me for this. I am afraid it will make it more difficult to act in the right way now – and please forgive me too if, knowing my own circumstances, I have been to blame before now.
Yours most sincerely
R.E.P.

Trinity College
November 19th 1889

My darling Rose
I am sorry to find from your letter that you have been exerting your unrivalled powers of making yourself miserable so successfully. I am writing this in case you have not come to the Laboratory, because I *must* see your father this afternoon and ask his consent to our marriage and he ought to hear of it first from you. I shall call at St Peter's Terrace to see your Father about 3. I will endeavour to reason with you when I see you again. In the meantime I am
Yours most affectionately
J.J.T.

Evidently the meeting at St Peter's Terrace went well and Sir George persuaded Rose that her scruples on his behalf were unnecessary. After all, he still had his wife to run the house as well as two other married daughters in Cambridge. J.J.'s forcefulness had won the day. No time was lost in planning the wedding and organising their future home.

Egerton Road, Fallowfield, Nr. Manchester
Dec. 24 1889

Dear Rose
I find that my ignorance of the ceremonials connected with marriage has led me into all sorts of solecisms with which I have been duly reproached by those who know. In particular I ought to have seen after the bridesmaid bouquet. If you will let me know what the exact colour is to be I will write to Solomon about it.

And now my own darling I have done with business and can indulge in the pleasure of wishing you every happiness in the Christmas which has now commenced. I hope that great as our love is now it will grow with each succeeding year and that in future years we may look back on this Christmas as the beginning of the happiest period of our lives. I will not attempt to tell you how much I love you. I know that my powers of expression are quite inadequate. I will ask you to take it for granted and conclude with the conventional wish. A Merry Christmas and a Happy *New Year*. Imagine I have given you a thousand kisses and send the same number to
Your own
J.J.

The academic families of Cambridge, as distinct from single men, were still numerically small in 1890 though the Statutes had been altered eight years earlier to allow all fellows, not just Heads of Colleges and Professors, to marry and retain their fellowships. J.J. as a bachelor living simply in College rooms was able after several years as a Professor to contemplate marriage. It is hardly surprising that he should have met Rose Paget. Her father was one of the most respected of the older professors and socially prominent, she was serious about physics and furthermore had quite striking looks; she was tall, rather dark with the curved 'Paget nose' which ran through the family, and deep-set blue eyes. She held herself very erect and the first impression was one of some severity, which was immediately softened when she spoke with a warmth and kindliness that made her friends at all levels. She was handsome in a memorable way rather than pretty. I lived with my grandmother during part of the Second World War when she was recently widowed and over eighty. Though the eyes were more deep set, the structure of her face was finely preserved and recognisable from photos of her at the age of thirty when she married.

Throughout her long life Rose always believed herself to suffer from poor health, mostly headaches and allergies. Travel was a tremendous business; even getting up to London was quite a major event. But a quiet life at home with the whole running of the household admirably managed by his wife suited J.J.'s own style. Domestic accounts were kept to the last halfpenny. It may be that a regular ordered life was the counterpoise J.J. needed to his incessant speculation about the universe and conceptions of matter; it was certainly the case that outside physics he was remarkably unadventurous in his activities and tastes.

Family life gave J.J. great pleasure. Though domestically he was quite unpractical and did no jobs around the house, except perhaps to arrange the wine for dinner parties, he was a devoted father. His son George was considerably in awe of him as a child and was in fact educated in the early years entirely by his mother. The relationship with his father grew closer as George began to show strong mathematical ability at the Perse School and himself came up to Trinity with a scholarship, as his father had, and took a double first in Maths and Physics. This led to an immediate offer of a fellowship at Corpus Christi College and the start of research at the Cavendish under his father. But the year was 1914 and at the outbreak of war, George, like most of his contemporaries, soon found himself commissioned into a foot regiment at the Front in France. There was no conception of holding back the brightest brains for scientific war work. I can remember my father telling me that of the ten undergraduate contemporaries present at the 21st birthday dinner given for him at home by his parents, only he and two others survived the war.

After the war George returned to Corpus and the Cavendish. It is never easy to work under a famous father in the same line of research, however close the relationship may be: anything good you do is bound to be seen as deriving from that link. George wisely put in for the professorship at Aberdeen

University while still in his twenties (a chair once held by James Clerk Maxwell) and pursued his own line with greater independence. This was to lead to a Nobel Prize for work which almost appeared to contradict his father's. Max Born has written about it succinctly:

> It is a fascinating fact that father and son have given the most striking evidence for the apparently contradictory properties of the electron: the father proving its character as a particle, the son its character as a wave. Modern quantum theory has shown that these are two aspects of the same phenomenon, depending on different kinds of observation; not contradicting but complementary. Thomson was extremely proud of his son's success and tried to assimilate the new results to his old convictions. [2]

Father and son collaborated on a major revision of J.J.'s *Conduction of Electricity through Gases* in the inter-war years. The relationship between them was remarkably free of any friction.

Joan was eleven years younger than George – the long gap being principally due to a stillborn child inbetween and Rose's always delicate state of health. As a consequence, Joan was brought up pretty much on her own, rather over-protected by her mother and increasingly close to her father, to whom she acted as secretary in his later days at Trinity Lodge.

As to recreations and holidays, these were all rather low-key. Unlike some other dons at Cambridge, he did not walk in the Lakes or sail on the Broads. As a young man he played a bit of lawn tennis, and also real tennis, and took the usual constitutional walks round Cambridge to nearby villages – Trumpington, Coton and Grantchester. Occasionally he took the train to Royston to play golf – usually by himself as it gave him the opportunity to think, and break off to look for wild flowers. In fact, flowers, particularly wild flowers, were his only other passion after physics. He used to say that if he could have had his life again he would have been a botanist, and that the discovery of how plants grew was as exciting as the discovery of how atoms were constructed.

He travelled on foot round Cambridge, and by train elsewhere. His son wrote, 'He could ride a bicycle but never used one, he never rode a horse, owned or drove a car, or travelled by aeroplane. He never went to a race meeting. His chief pleasures were in congenial talk – he enjoyed it especially when combined with a well cooked meal – his garden, watching Rugby football, tennis (real or lawn) or cricket and reading either detective stories or his favourite novels.' Dickens, Jane Austen and the Waverley novels of Scott were among his favourites. I can remember as a boy of seven, staying at Trinity Lodge, being given a novel by my grandmother, which she first carefully dusted, to take down to J.J. in his study. I remember this tiny episode for two reasons: first, the name of the book was *Martin Chuzzlewit* which I thought was a very odd name to have, and secondly because my grandfather was obviously at prayer in his study when I went in. Though in his early life he was regarded by his friends as a doubter, he was at heart in his private life a humble and quite devout man without pretension. He was also a traditionalist. As the first Master of Trinity who was not a clergyman, he would have been concerned that his appointment should not be seen as a break with the past, particularly in such public matters as attending chapel, which he occasionally did before he was Master, and regularly thereafter.

By the early 1890s J.J. had come a long way from his Manchester roots. He was a Professorial Fellow of Trinity. He had played himself in carefully at the Cavendish and, while continuing the earlier lines of work and teaching of his predecessor, had found for himself a major subject of research in the conduction of electricity through gases. Within a few years, this was to turn into a winner, leading him personally to his greatest discoveries and bringing a world-wide reputation to the laboratory. He had married happily into one of the best-known Cambridge families of the time, and Rose had given birth to a son. He was a star in the ascendant, attracting the brightest students to the Cavendish.

The growth of the Cavendish was also forwarded, fortuitously but happily, by the University's institution of research degrees for students coming from outside. At the same time, a change in the regulations of the 1851 Exhibition Commission allowed that wealthy body to fund science research scholarships for overseas students. The first of these was Ernest Rutherford; while still at home he had read everything that J.J. had written and decided that this was the man under whom he wished to work. It was the start of a long and fruitful collaboration and friendship. In a photograph of the Research Students at the Cavendish taken in 1898, as many as nine out of sixteen were holders of these research scholarships from outside Cambridge.

It was the magnetic quality of J.J. that drew in so many of these brightest minds, and formed a seed-bed that was continually renewed as established members went off to professorships around the world (75 of them altogether at 55 universities) and new students arrived. For more than twenty years this group of research workers, of which J.J. was the head, was easily the most important of any physics laboratory in Britain, and probably in the world. It included a formidable list of scientists, both British and foreign, who in their turn were to make fundamental contributions to science: at least eight of his pupils won Nobel Prizes.

What was the essence of J.J.'s greatness as a scientist? How was it that J.J. grasped the underlying significance of an experiment in a way that to some seemed almost uncanny? How did he relate theory to experiment and how did he create such an atmosphere of creative excitement among his researchers?

My father, who of course collaborated with his own father closely, put it like this:

> In all his theories J.J. liked to visualise, and for him the mathematics was always merely the language which described the physical and spatial concepts in his mind. He had no idea of mathematics dictating the theory. [3]

It is well known that he was not personally good at the physical handling of apparatus; he devised it in his mind and Ebeneezer Everett, his personal assistant for most of his time at the Cavendish, constructed it and manipulated it into working. To quote my father again:

> He had the physician's gift of diagnosis and could often tell a research worker what was really the matter with an apparatus that a man had made and struggled with miserably for weeks. His own apparatus was simply designed and constructed without unnecessary refinement. The phrase 'sealing wax and string' with which a later generation described the Cavendish apparatus of his day is an exaggeration of course but not a great one, judged by modern standards at least there was a slightly amateurish air about it . . . Yet this rather odd collection of glass and brass did in fact play a major part in producing a change in men's ideas of the nature of matter and energy that has occurred since 1896.

J.J.'s own mind showed 'restless mental activity and originality', as one former student, Professor John Zeeman, said of him in the salutations that marked J.J.'s 70th birthday. The subjects he pursued were for him completely the most exciting things in the world, and he communicated this excitement to his researchers. What he valued most in research was enthusiasm, even more than originality, though he valued that very highly.

> Though most of the people in the Cavendish in his day were working more or less on the same lines, working on the implications of the new discoveries, X-rays and electrons, especially as they concerned the conduction of electricity in gases, there was no attempt to direct research and quite a few were working on other lines. He did not believe in too much direction. [4]

Also he had a very human side to him which drew forth a remarkable warmth of feeling and affection from those who worked under him:

> Those of us who have been more fortunate and to be privileged to associate with him at one time or another have had an increasing esteem for him not only because of the great scientific achievement which his mind made possible, but also for those things which a man likes to find in another man. Not only did he show interest in one's scientific work, but also he always showed human interest in the man himself, his friends, and the institutions with which he was connected. We all love his characteristic smile, and every one of us felt a certain pleasure within ourselves on hearing a footstep that every Cavendish man recognises as solely J.J.'s. [5]

J.J. could be forceful and determined in getting what he wanted for his own work for the laboratory and for his students, and was always conscious that he possessed rare intellectual power, yet he was modest about his own achievements. Throughout his written work he took close account of the contribution of others. This showed particularly in the field of cathode rays where much work had been done by German physicists (indeed the name Kathoden-Strahlen originated in Germany) such as Plucker, von Helmholtz and his famous pupil Hertz, Wiedemann, Goldstein, Lenard, Kaufmann and Wiechert, and by Perrin in France. In the great upsurge of activity following

Röntgen's discovery of X-rays there was a good deal of parallel activity taking place and no doubt J.J.'s great discovery that the cathode ray stream was made up of particles with a mass of a smaller order of magnitude than an atom, and which were universal constituents of matter, would have been postulated before long by others in this field who were conducting very similar experiments. But the fact is he was the first to see the profound significance of these experiments and it took several years to convince others that he was right. J.J. learned enough German, with the help of Rose who spoke it fluently, to read the German scientific journals and kept closely abreast of the work going on elsewhere, not only through the journals but also by direct correspondence.

As to his honours other than the Nobel Prize, he held the main awards given by the Royal Society, in particular the Copley Medal, and a large number (I think 23) honorary doctorates at other universities. Membership of the Order of Merit meant much to him. But it gave him no more pleasure than becoming Master of Trinity, the college which was for him in personal terms so close to his heart. He declined a peerage partly on the grounds, my father told me, that he did not think himself wealthy enough to carry the honour, though by the end of his life he certainly would have been. In fact he died quite a wealthy man but this was through shrewdness on the Stock Exchange and not through his exploitation of any of his discoveries. He never took out any patents (as Lord Kelvin for example did) though the cathode ray oscilloscope and the television tube derive directly from his apparatus, and in a broader sense the whole of today's electronics industry descends from his breakthrough. At the time of J.J.'s key discovery of the electron, no one could have foreseen the vast range of its practical applications a century later. He was always interested in the commercial applications but did not seek to benefit from them personally.

My father summed up his own memories of J.J.:

> Throughout his life J.J. was always aware of the over-riding importance of intangible things, family love, friendships, kindness and sincere religion. This awareness that shone through all he did, made him memorable as a man as well as a physicist. [4]

References

[1] Crowther, J. G., *The Cavendish Laboratory 1874–1974*, 1974, London: Macmillan, p. 103.

[2] Born, M., Obituary Notice of Sir J. J. Thomson. *Proceedings of the Physical Society*, **53** iii (1941).

[3] Thomson, G. P., Manuscript autobiography. Trinity College Library, Cambridge.

[4] Thomson, G. P. and Thomson J., J. J. Thomson as we remember him. *Notes and Records of the Royal Society*, **12** ii (December 1957).

[5] Professor Alois F. Kovarick quoted in Rayleigh, fourth Lord, *The Life of Sir J. J. Thomson*, 1942, Cambridge University Press. Reprinted in 1969, London: Dawsons, p. 228.

Preface

In this book we aim to give a readable account of J.J. Thomson's work on the electron, in the context of his life and other scientific work. We also hope to give enough background to help the reader interpret the original papers which are reproduced at the ends of Chapters 2–6. These are important reading for anyone who wants to get an insight into Thomson and his way of thinking, besides being historically significant in themselves.

However, this approach has its limitations, which we enter here. This is not a biography. Thomson was a brilliant man, both personally and intellectually, and the electron work was just one episode in a many-faceted career. Nor is this a full account of the discovery of the electron. Many people, now largely forgotten, contributed to the idea of the electron. While we mention some, we do not list all of them, and do not discuss their work in any detail. To many German physicists in the early nineteenth century, the contributions of Lorentz, Zeeman, Wiechert and Kaufmann, for example, were seen as of major importance. We hope that some day their story will also be told.

The authors wish to thank the Cavendish Laboratory, University of Cambridge, for permission to reproduce the Plates on pp. 21, 22, 49, 57, 58, 81, 82, 115, 116, 127, 128, 197, 198, 201, 202, 207 and 213, and Keith Papworth for supplying them. Cambridge University Library is acknowledged for providing access to original manuscripts and for permission to reproduce the Plate on p. iii. The Plates on pp. ii, 2, 12, 50, 117, 118, 208, 213 and 214 are reproduced from original photographs owned by J.J.'s grandson, David Thomson, to whom we are grateful for their loan, for his cooperation generally and most of all for writing the illuminating Foreword to this book. Finally we thank Andrew Carrick, David Courtney, Sarah Ramirez and Maureen Allen for their assistance with production and Stephanie Brooks for typing several sections of the manuscript.

1

Formative Years

J.J. Thomson, best known today as the 'discoverer of the electron' was an important and pivotal figure in the history of British physics. His 35-year-long professorship at the Cavendish Laboratory in Cambridge saw the Laboratory, and the generation of physicists he educated, break with nineteenth- century physical methods and concerns and become world leaders in the new twentieth-century physics with its concern for microphysics – elementary particles, atomic theory, the structure of light, etc. Thomson's own work marks the break with previous tradition, although his fundamental concepts remained rooted in his nineteenth-century background. To understand what he was doing when he 'discovered the electron' we have to look closely at his education and previous work, at how the electron work developed from his earlier studies, and at how it opened up new areas of physics.

Joseph John Thomson was born on 18 December 1856 at Cheetham Hill, a suburb of Manchester. His father, Joseph James Thomson was an antiquarian bookseller, a business which had been in the family for three generations. Thomson believed that the family came originally from lowland Scotland. However, by the time he was born they were well established in the Manchester area and had wide connections, largely in the textile industry. Thomson's mother, Emma Swindells, came from a branch of the local Vernon family who owned a cotton spinning company, and Thomson had a network of cousins around Manchester. His only brother, Frederick Vernon Thomson, was two years younger and later went into business with Claflin & Co., calico merchants. The family seems to have been a very close and united one, largely occupied with helping with parochial church work, and both parents clearly put considerable thought into their sons' education. Thomson's early interests in science and gardening were encouraged. After his father's death his mother and brother both made considerable sacrifices to see

Thomson as a child

Thomson through college, and both sons remained devoted to their mother, spending their summer holidays with her until her death in 1902. In 1914, when Frederick Vernon was seriously ill following an operation, he moved to Cambridge to be near J.J., and lived there the remaining three years of his life.

Education at Owens College

J.J. Thomson's early education was at small private schools where he learnt the rules of Latin grammar, passages of English literature and the propositions of Euclid, all by heart. Then came what he considered to be the turning point of his life. He was sent to Owens College almost, it seems, by accident, as he later recalled:

> . . . it was intended that I should be an engineer . . . It was arranged that I should be apprenticed to Sharp-Stewart & Co., who had a great reputation as makers of locomotives, but they told my father that they had a long waiting list, and it would be some time before I could begin work. My father happened to mention this to a friend, who said, 'If I were you instead of leaving the boy at school I should send him while he is waiting to the Owens College: it must be a pretty good kind of place, for young John Hopkinson who has just come out Senior Wrangler at Cambridge was educated there'. My father took this advice, and I went to the College. [1]

Thomson's reminiscence highlights the supreme position of Cambridge in British mathematical education at the time, with an influence extending far beyond the University, and even beyond mathematics, for the Mathematics Tripos was considered as training for the mind within a liberal education. Through the nineteenth century the Mathematics Tripos became increasingly competitive and to be a high Wrangler (as the first-class degree men were called) was a supreme intellectual distinction in Victorian Britain. The name of the Senior Wrangler (the top man) was published in all national newspapers, postcards of him were on sale, and his home town and old school would celebrate with torchlit processions and a day's holiday.

Manchester in the mid-nineteenth century was a rapidly growing, bustling and energetic city noted not only for being the hub of Britain's vital cotton industry, but also as a centre of invention and progress. Hepworth Dixon echoed the sentiments of Mancunians when he wrote in the *People's Journal* in 1847, 'Invention, physical progress, discovery are the war-cries of today. Of this great movement Manchester is the centre. In that lies its especial importance. That work which it seems the destiny of the nineteenth century to accomplish is there being done.' [2]

In this industrial culture the potential benefits of science for technology were recognised and institutions were established to promote and exploit science. Civic pride required that the city's cultural resources be developed to match its economic status. Owens College was the outcome of these concerns. It was financed by John Owens, a wealthy Manchester merchant who, in 1846, left a bequest for the instruction of young men 'in such branches of learning and sciences as now and may be hereafter taught in the English Universities'. [3] Owens College was modelled on the Scottish universities and the colleges of London University, to which it affiliated. Professors were appointed in classics, mathematics and natural philosophy (physics), mental and moral philosophy, English, chemistry, natural history, German and French, the last four chairs being half-time. A house was taken and converted in Quay Street, and the College opened in 1851.

As Thomson pointed out, of his admission to Owens, such good fortune could not have happened at any other place or time. Manchester was the only provincial city which had anything corresponding to Owens College. Furthermore, 'I was only fourteen when I went to Owens College, and I believe that sixteen is now the minimum age at which students are admitted to such Colleges. Indeed the authorities at Owens College thought my admission such a scandal – I expect they feared that students would soon be coming in perambulators – that they passed regulations raising the minimum age for admission, so that such a catastrophe should not happen again'. [4] This is the first positive indication we have of Thomson's brilliance, although it had clearly been evident to his classmates earlier on. One, who achieved prominence in local government, later boasted in *The Manchester Guardian* that, 'There was a clever boy at school with me, little Joey Thomson, who took all the prizes. But what good has all his book learning done him? Who ever hears of little Joey Thomson now?' (This was later quoted to Thomson in an after-dinner speech. J.J., replying, turned the joke, 'I wish you were not going to hear little Joey Thomson now'.) [5].

Two years after he entered Owens, in 1873, Thomson's father died, leaving the family poorly off. Thomson's brother, Frederick, left school and got a job to supplement their income. They could no longer afford the cost of an engineering apprenticeship for Thomson and he was left to make his way by scholarships, thus forcing him to concentrate on those subjects he was good at: mathematics and physics.

At Owens, Thomson came under the influence of three men in particular: Thomas Barker (Professor of Mathematics), Balfour Stewart (Professor of Natural Philosophy) and Osborne Reynolds (Professor of Engineering). From them he acquired most of the attitudes and images which were to form his intellectual repertoire throughout his life.

Towards the end of his life Thomson recalled Barker as the best teacher of mathematics he had ever come across. His lectures were always carefully prepared and he aimed to give his pupils a sound grounding in the fundamentals of mathematics rather than training them to solve problems rapidly. Barker was himself a Cambridge senior wrangler, who extended his teaching far beyond the usual schoolboy range of arithmetic and Euclid. Thus Thomson received lectures on, for example, the logic of mathematics, and on quaternions, a system of geometrical analysis introduced by Sir William Rowan

Hamilton, in which there was a revival of interest at the time. In the light of later experience Thomson rejected quaternions, judging that, 'Though the ideas introduced by Hamilton were very interesting and attractive, and though many physical laws, notably those of electrodynamics, are most concisely expressed in Quaternionic Notation, I always found that to solve a new problem in mathematical physics, unless it was of the simplest character, the older methods were more manageable and efficient.' [6] However, he did think Barker's use of such subjects to stimulate interest successful: 'I do not suppose that the introduction of unusual subjects such as these in comparatively elementary courses is always to be recommended, but I think it was successful in this case and that we worked all the better for it; we regarded ourselves as pioneers and that it was up to us to make good.' [7] Thus, even before entering Cambridge, Thomson's mathematical education was geared to the requirements of the Mathematics Tripos.

Balfour Stewart, Professor of Natural Philosophy, started his career as assistant to Professor David Forbes in Edinburgh, where he made his name by his research on heat. He later became Director of the Kew Observatory before coming to Owens College in 1870. Like Barker, Balfour Stewart was an inspiring teacher and very enthusiastic about research. Arthur Schuster, also a pupil of his, remarked that Stewart was fond of arguing by analogy, and also that he often tried refined experiments with inadequate equipment, a trait which Thomson may well have picked up from him. [8] One such experiment was on trying to detect whether there was a change in weight on chemical combination, one of a series on the fundamentals of gravitation. By this time Thomson had become inspired with Stewart's enthusiasm and was helping him with these experiments:

> One Saturday afternoon . . . when I was alone in the laboratory, after tilting the flask, though the mercury ran over the iodine [the combination of which was being tested] no combination took place. I held it up before my face to see what was the matter, when it suddenly exploded; the hot compound of mercury and iodine went over my face and pieces of glass flew into my eyes. I managed to get out of the laboratory and found a porter, who summoned a doctor. For some days it was doubtful whether I should recover my sight. Mercifully I did so, and was able after a few weeks to get to work again. Modern physics teaches that a change of weight is produced by chemical combination, and enables us to predict what it should be. In the experiment I was trying, the change of weight would not be as great as one part in many thousand millions, so that it is not surprising that I did not detect it. [9]

Thomson admired Stewart for his vigorous mind and the encouragement he gave his students to work out their own ideas, both theoretical and experimental. Thomson later emulated Stewart in these respects and in his willingness to publish his working hypotheses, however tentative and ephemeral these might be.

From Stewart, Thomson received a thorough grounding in the prevalent Victorian method of reasoning by analogy, and in ether physics. Belief in the ultimate unity of knowledge was strong in the late nineteenth century. Unification was sought in the ether, which was thought to be some sort of subtle fluid whose structure and dynamics would explain all phenomena. Thomson, whose later Cambridge education reinforced his belief in the ether and in physical analogies and added to them the mathematical methods of dynamics, believed that all phenomena, even atoms themselves, could ultimately be explained as constructed of a fluid ether. He was prepared to do without matter on occasion, but never without the ether. The ether unified his views of matter, electricity and energy in a complex relationship: both matter and electricity were some sort of structure of the ether, while energy represented the unseen motion of these structures.

Thomson's preferred analogies were all of vortices in the ether. He started his career by investigating the possibility that atoms might be vortex rings (like smoke rings) in the ether. He later abandoned this idea, using ethereal vortices as an explanation of electricity and the electromagnetic field instead. As late as 1931, long after ether physics had become unfashionable and explicit mention of it had dropped out of his publications, he wrote to Oliver Lodge, 'I agree with you about the close connection between electricity and vortex motion. I have always pictured a line of electric force as a vortex filament . . . ' [10]

One of the origins of Thomson's fascination with vortex motion lay with a third teacher, Osborne Reynolds, Professor of Engineering at Owens. Reynolds was, according to Thomson,

> . . . one of the most original and independent of men, and never did anything or expressed himself like anybody else. The result was that it was very difficult to take notes at his lectures, so that we had to trust mainly to Rankine's textbooks . . . He would come rushing into the room pulling on his gown as he came through the door, take a volume of Rankine from the table, open it apparently at random, see some formula or other and say it was wrong. He then went up to the blackboard to prove this. He wrote on the board with his back to us, talking to himself, and every now and then rubbed it all out and said that was wrong. He would then start afresh on a new line, and so on. Generally, towards the end of the lecture, he would finish one which he did not rub out, and say that this proved that Rankine was right after all. [11]

Reynolds' lectures must have been an unmitigated disaster for most students. Thomson, however, clearly benefited both from the experience of seeing, 'an acute mind grappling with a new problem' [12] and from the understanding of Rankine's ideas that he acquired. Rankine and Reynolds both believed that engineering should be taught as a science, with an emphasis on fundamental theory. Rankine, in his textbooks, used vortex models extensively to explain thermodynamics.

Equally important to Thomson were Reynolds' own research experiments in the 1870s, on vortices. Investigating turbulence (the Reynolds number is named after him), Reynolds injected dyes into the water behind moving discs. The dyes made the resultant vortex rings clearly visible. These experiments probably made a great impression on Thomson, who had a strongly visual imagination and recalled them as 'very beautiful'. [13] They gave him a visual idea of the behaviour of vortices and an intuitive understanding of them which lasted the rest of his life. Like Reynolds, Thomson, in his later research, tended to form an idea of what was happening before he analysed it mathematically. For Thomson these ideas were visual. His surviving working notes nearly always take the form of sketches, generally of vortices in some form, which represent the central idea, surrounded by a halo of equations elucidating details. An example is shown in Figure 4.1 on p. 78.

Shortly before he left Owens, Thomson made two further contacts which were to prove influential. The first was with Arthur Schuster, who took up a demonstratorship at Owens in 1873. Schuster gave a course based on Maxwell's *Treatise on Electricity and Magnetism* which initiated Thomson's devotion to Maxwell's theory, later fostered by William Niven's lectures at Cambridge. Unlike many of his contemporaries, Thomson based his understanding of Maxwell not only on the *Treatise*, but also on the earlier papers, which were themselves based on vortex analogies. Thomson was so fascinated by these papers when he first saw them that he copied them out long-hand; no light task for they are long papers.

The second contact was with John Henry Poynting who returned to Manchester in 1876 after graduating at Cambridge as Third Wrangler (third top). Poynting became a great interpreter of Maxwell's theory and many of Thomson's later views of electromagnetic theory were based on Poynting's developments. The two men later co-authored a well-known series of textbooks, mostly written by Poynting. Their friendship, which lasted until Poynting's death in 1914, Thomson counted 'one of the greatest joys of my life'. [14]

By this time Thomson had given up thoughts of engineering. He had a great ambition to do original research and before he left Owens, published his own first paper, 'On Contact Electricity of Insulators', an experimental work elucidating a detail of Maxwell's theory. Thomson's wide ambitions are indicated in his publishing this in the *Proceedings of the Royal Society*, rather than in any local, but easier journal. They are also indicated in his trying, at Barker's encouragement, for a mathematical scholarship at Trinity College, Cambridge, the most prestigious and difficult college to get into. He tried twice, once in 1875, clearly before he was really ready, and again in 1876 when he won a minor scholarship and a subsizarship (a student who paid reduced fees in return for undertaking one or two minor college responsibilities).

By the time he went to Cambridge at the age of 20, Thomson had gained from Owens the major stylistic themes of his later work: the familiarity with advanced mathematics, the use of vortex analogies within an ether-based physics, a willingness to advance speculative theories, and enthusiasm for research.

These were to be refined, but not undermined, by his intensive Cambridge training in the methods of analytical dynamics.

Cambridge and the Mathematics Tripos

Thomson went up to Trinity in the autumn of 1876, and essentially remained there until his death 64 years later in 1940, priding himself on having 'kept' every term. This, in itself, indicates his devotion to his college and to Cambridge, where he felt immediately at home. He was known to claim that the climate of Cambridge suited him and he felt well and vigorous when there.

At Cambridge, Thomson became absorbed into the world of the Mathematical Tripos, and particularly of those aspiring to a high place in it. In the late nineteenth century the Mathematical Tripos was a major focus of intellectual challenge and enormous kudos attached to achieving a high result. It had developed customs, traditions and a way of life which were all its own.

The examination itself, generally taken after three years, consisted of 15–20 three-hour exams, sat over a period of three weeks. A few days later the results were read out in order of merit, men first, from the gallery of a crowded Senate House, after which printed class lists were tossed down, to be fought for by the waiting crowd. Students were divided into Wranglers (first class), Senior Optimes (second class) and Junior Optimes (third class). The Senior Wrangler (top man) was feted throughout Cambridge, by his school and in his home town. The unfortunate man who came last (the bottom pass) later had a ceremonial wooden spoon, ornately decorated and generally man-size, presented to him in the Senate House after he had taken his degree.

Aspiring Wranglers took their training extremely seriously, working intensively for eight hours each day, taking two hours exercise, generally in the form of long fast walks in pairs, relaxing after dinner, and keeping themselves celibate. The strain of the Tripos exam itself was immense. Thomson recorded that he suffered badly from insomnia during it, but relaxed himself with a shampoo at the barbers each day between morning and afternoon papers.

The philosophy behind the Mathematics Tripos was that of a liberal education as interpreted by William Whewell. Whewell distinguished between 'established knowledge' and 'research'. The former was stable and well suited to training the mind, especially when cast in a form that generated a range of well-defined problems and solutions – Euclid's geometry and Newton's mechanics were ideal. These were presented in a way that kept the mathematics grounded in reality – in pulleys, machines, forces, etc. – but in an idealised form only. Students were certainly not encouraged to experiment and get 'hands on' experience of these realities.

By the 1870s analysis was central to the syllabus in the form of analytical geometry and dynamics. To be a successful analyst, however, took years of intensive training and a system of coaching had grown up to prepare aspiring mathematicians for the increasingly competitive Tripos exam. Thomson's coach, Edward Routh, was far and away the most important influence on his mathematical thinking. Routh was the most famous and successful of all Cambridge coaches. He had been Senior Wrangler in 1854, his original research was in analytical dynamics and he grounded his students thoroughly in its methods. Analytical dynamics, i.e. the use of Lagrange's equations and Hamilton's principle of varying action, was a specific mathematical formulation which embodied the mechanical philosophy, the belief that all phenomena could ultimately be explained on mechanical principles in terms of some sort of bodies in motion. This was well entrenched in British mathematics and physics in the 1870s. The power of the method lay in its ability to coordinate and predict phenomena by analysing the energy changes involved. Knowledge of the underlying mechanism was unnecessary. Thomson later likened it to a watch or clock, where one can, and commonly does, know how the movements of the hour and minute hands are related, without having any idea of the mechanism behind the clock face.

Routh's teaching covered, as Thomson recalled, 'practically all the branches of pure and applied mathematics known at the time'. [15] The main physical subjects were statics, dynamics, hydrostatics, optics and astronomy, Newton on planetary motion, and electricity and magnetism in the analytical mathematical form developed by William Thomson (later Lord Kelvin and no relation of J.J., hereafter referred to as Kelvin) and James Clerk Maxwell. However, even in 1881 when Thomson took the Tripos, and nine years after the founding of the University Physics Department, the Cavendish Laboratory, students were still discouraged from relating mathematical theory directly to experiment.

Many students considered College and University lectures irrelevant to passing the Tripos, relying solely on their coaches. But Thomson attended lectures given by the College Lecturers William Niven and James Glaisher, and by University Professors Arthur Cayley, John Couch Adams and George Gabriel Stokes. This indicates again his impatience with the merely routine, and with the repetitious exercises set by the coaches to drive methods home and improve their pupils' speed in the Tripos Exam, as well as his desire to seek out the most exciting and up-to-date branches of his subject.

The lectures by Niven and Glaisher had the greatest long-term influence on him. Glaisher's were obviously fascinating and he was the only person at Cambridge to foster Thomson's interest in research, encouraging him to publish three mathematical papers in the *Messenger of Mathematics*. Niven, in contrast, was an extremely bad lecturer and once again, as with Reynolds, Thomson was driven to reading through the original books for elucidation. The subject was electromagnetism and the books were Maxwell's *Treatise* and the related papers, reinforcing Thomson's earlier acquaintance with these. However, Niven made up in enthusiasm for Maxwell's theory what he lacked as a lecturer and, based on their mutual interest, he became 'one of the best and kindest friends I [Thomson] ever had'. [16] He introduced Thomson to a number of College Fellows and later was on both the appointment committees which elected Thomson first to a college fellowship, and later to the Cavendish Professorship, urging his merits on each occasion. Niven later became godfather to Thomson's son, George.

Thomson graduated as Second Wrangler in 1880, Joseph Larmor being the First. (Larmor was to become Lucasian Professor of Mathematics at Cambridge and originated much of the theory of electrons.) Thomson's training in the Mathematics Tripos had added a gloss to his earlier, Owens, conceptions. He still sought unification of knowledge by the methods of analogies within an ether-based physics, but he had had no contact with experiment for three years. He now had a thorough grounding in mathematics and analytical dynamics to help him analyse his analogies. He was well established in Cambridge where he had achieved great success.

References

[1] THOMSON, J.J., *Recollections and Reflections*, 1936, London: Bell. Reprinted 1975, New York: Arno, pp. 1–2. Page references given are to the later edition.
[2] DIXON, H., *People's Journal*, **3** (1847), 246.
[3] Quoted in KARGON, R., *Science in Victorian Manchester*, 1977, Manchester: Manchester University Press, p. 155.
[4] See reference [1], p. 2.
[5] RAYLEIGH, FOURTH LORD, *The Life of Sir J.J. Thomson*, 1942, Cambridge University Press. Reprinted 1969, London: Dawsons, p. 269.
[6] See reference [1], p. 14.
[7] See reference [1], p. 14.
[8] SCHUSTER, A., *Biographical Fragments*, 1932, London: Macmillan, p. 213.
[9] See reference [1], pp. 20–21.
[10] Birmingham University Library, Lodge papers, OJL 1/404/38.
[11] See reference [1], p. 15.
[12] See reference [1], p. 15.
[13] See reference [1], p. 17.
[14] See reference [1], p. 22.
[15] See reference [1], p. 36.
[16] See reference [1], p. 43.

2

Early Research

Analytical Dynamics

For the first four years after graduating, Thomson's work was dominated by his commitment to analytical dynamics and to Maxwell's electrodynamics. Having no independent means, and aiming high as always, he tried for a fellowship at Trinity the first year after he graduated, rather than waiting the customary two. Contrary to the prognostications of his friends, who had advised him not to be so ambitious, he got the fellowship. For his fellowship thesis he returned to an idea he first conceived while attending Balfour Stewart's lectures on the conservation of energy:

> Stewart in his lectures paid special attention to the principle of the Conservation of Energy, and gave a course of lectures entirely on this subject, and naturally I puzzled my head a great deal about it, especially about the transformation of one kind of energy into another – kinetic energy into potential energy, for example. I found the idea of kinetic energy being transformed into something of quite a different nature very perplexing, and it seemed to me simpler to suppose that all energy was of the same kind, and that the 'transformation' of energy could be more correctly described as the transference of kinetic energy from one home to another, the effects it produced depending on the nature of its home. This had been recognised in the case of the transformation of the kinetic energy of a moving body striking against a target into heat, the energy of the heated body being the kinetic energy of its molecules, and it seemed to me that the same thing might apply to other kinds of energy. One day I plucked up courage to bring this view before Stewart. I should not have been surprised if he had regarded me as a heretic of the worst kind, and upbraided me for having profited so little from his teaching. He was, however, quite sympathetic. He did not profess to agree with it,

J.J. Thomson, around 1880

> but thought it was not altogether irrational, and that it might be worth my while to develop it. [1]

Now, four or five years later, and armed with the methods of analytical dynamics, Thomson investigated the transformation of energy using a modified Lagrangian formulated only four years previously by Routh in his Adams Prize Essay of 1877. The essential idea of the thesis was that all types of energy were manifestations of the kinetic energy of invisible systems.

In this work he also evidenced, for the first time, a belief in the physics of the unseen common to many late nineteenth-century physicists, and in particular to Balfour Stewart. In 1875, Stewart and Peter Guthrie Tait, Professor of Natural Philosophy at Edinburgh, had published *The Unseen Universe*, a refutation of Tyndall's Belfast Address of 1874 in particular, and of nineteenth-century naturalism in general. Stewart and Tait attempted to show that there was a place for God in physics: natural order included an invisible realm which communicated with the visible realm. Divine intervention, acting under the operation of natural laws, transferred energy from the visible to the invisible realm and vice versa. While there is no explicit indication that Thomson's belief in the unseen had any religious overtones, this is the tradition in which he was educated.

Thomson's fellowship thesis was never published, but he developed the ideas further in two long papers in the *Philosophical Transactions of the Royal Society*, and in a book, published in 1888, *Applications of Dynamics to Physics and Chemistry*, which analyses and correlates in detail the energy changes in a whole range of different phenomena from simple situations of moving bodies, to changes of state, and chemical reactions.

Thomson's extreme preference for dynamical methods, over those of thermodynamics, is an indication of his commitment both to the mechanical philosophy, and also to unifying physics by correlating phenomena. However, he was also beginning to realise the limitations of the method: analytical dynamical results are couched in terms of theoretical quantities, such as energy, rather than experimental observables, such as temperature. Knowledge, or some sort of model of the system, is required to translate the theoretical prediction into an experimentally testable hypothesis. An example from an early, unpublished, thesis shows this: experiments by Lengstrom had shown that a ring of insulating material rotating at high speed could act like an electric current. Thomson applied analytical dynamics to this result and predicted that if the experiment was performed with an electrically charged ring, then the motion of the molecules within the ring would alter as the speed of rotation increased. He hypothesised that this altered motion would manifest itself as a change in the temperature of the ring. But since he lacked a model of exactly how the experimentally observable temperature change was related to the unobservable motion of the molecules, he could not predict the magnitude of the temperature change, thus making good experimental design and testing extremely difficult.

This was a constant weakness of analytical dynamics. It could predict one phenomenon from another but, without a detailed model of the mechanism involved, it gave no indication of the size of effect to be looked for. This made direct quantitative comparison between theory and experiment virtually impossible. Thomson's realisation of this limitation had a profound effect on his attitude to experiment. Except for a few experiments performed around this time (in the early 1880s), he ceased to try to perform experiments in which accurate measurement tested detailed theoretical predictions. (Indeed, for a year or two he virtually stopped experimenting, publishing theoretical papers which relied on other people's experimental results for supporting evidence.) The situation was made worse by his own mathematical investigations into analytical dynamics. In his work on potential and kinetic energy, Thomson demonstrated that potential energy was formally equivalent to the kinetic energy of a system characterised by cyclic coordinates. The essential property of such a system was that its motion was unseen as Thomson, following Stewart and Tait, required. But he also showed that since the mechanism was cyclic and unseen, the Lagrangian formulation could not tell him anything about it. Thus, while he had demonstrated that a mechanical explanation of potential energy was possible, Thomson had also revealed a block in the method of analytical dynamics: such methods could not discover what the unseen mechanism was.

Thomson's work on analytical dynamics reinforced his idea that theories should coordinate phenomena, and that the best way to formulate a theory was to gather together different experimental phenomena. At the same time, however, theories of underlying mechanisms could never be known to be 'true'; they were at best heuristic tools. Consequently, such theories need not correspond exactly to precise experiment. Thomson found this idea theoretically liberating. He felt free to use more than one model of the same mechanism at the same time, or to flit rapidly between suggested mechanisms, since none were the real mechanism. This was a philosophy which suited him admirably, and he became renowned for the fertility of his theoretical imagination.

Electromagnetic Mass

Upon his election to a fellowship, Thomson began mathematical investigations on moving charges of electricity according to Maxwell's theory. In 'On the Electric and Magnetic Effects Produced by the Motion of Electrified Bodies', published in the *Philosophical Magazine* in 1881, he considered the case of a charged sphere moving in a straight line and reasoned that the changing electric field in its neighbourhood must produce a magnetic field and hence endow the surroundings with energy. The mass of the charged sphere is thereby effectively increased over and above the mass it would have if uncharged. Thomson drew an analogy with a sphere moving through water; the sphere cannot move without setting the surrounding water in motion and hence its mass is enhanced by that of the moving water. Part of this paper is reproduced on p. 25.

The idea that the mass of a body might be partly electrical in origin was completely novel. Thomson did not, however, consider that the concept undermined Newtonian mechanics based on the idea of inertial mass. The extra, electromagnetic, mass resided in the ether surrounding the body, not in the body itself. His view at the time was of a strained ether. Later he regarded the extra mass as residing in the Faraday tubes which radiated from the charged particle and, still later, after he had identified cathode rays with corpuscles (see Chapter 5), he considered the possibility that the mass of these corpuscles was entirely electromagnetic in origin. He and others developed this idea again after 1900, basing the entire corpuscular theory of matter on it.

In 1964, G.P. Thomson wrote in his biography of J.J.:

> This idea of electromagnetic mass was the first hint of a connection between mass and energy, although J.J. did not put it in that way till later. A quarter of a century later, Einstein made this connection the basis of his famous equation $E = mc^2$, stating the equivalence of mass (m) and energy (E). In fact J.J. got the numerical factor wrong, and this may well have prevented him from seeing the connection. [2]

Thomson's theory was later developed and corrected by many others beginning with George FitzGerald and Oliver Heaviside.

College Lectureship – Vortex Atoms

Around 1882, about the time he finished his paper on electrified particles, Thomson was elected to an Assistant Lectureship in mathematics at Trinity College. As a fellow, and then a college lecturer, Thomson lived in college and devoted a large part of his time to college teaching in mathematics. This took up to 18 hours a week, but he always claimed to have enjoyed and benefited from it. The assistant lectureship to which he was appointed seems to have been part of an attempt by the college to regain control from coaches, such as Routh, and marks the beginning of Cambridge's supervision (tutorial) system:

> At this time the College was taking steps to remove the anomaly – to call it by no harsher name – that while all the candidates for the Mathematical Tripos paid fees to the College for lectures in mathematics, whether they attended them or not, they did all they could to avoid going to these lectures, and relied on private tuition for their teaching. The system proposed by the College was that the College Lecturers, in addition to giving lectures, should take men individually, help them with their difficulties and advise them as to their reading. I was elected on the understanding that I should take part in this work. [3]

However, college teaching did not distract him from research. Cambridge University's Adams Prize was awarded every two years for an essay in some branch of pure maths, astronomy or natural philosophy. The subject was set in advance and was generally of topical interest. In 1882 the subject set was 'A general investigation of the action upon each other of two closed vortices in a perfect incompressible fluid'. As Thomson recalled, this was a subject that immediately appealed to him:

> I was greatly interested in vortex motion since Sir William Thomson had suggested that matter might be made up of vortex rings in a perfect fluid [the ether], a theory more fundamental and definite than any that had been advanced before. There was a spartan simplicity about it. The material of the universe was an incompressible perfect fluid and all the properties of matter were due to the motion of this fluid . . . The investigation of the problem set for the Adams Prize, like most problems in vortex motion, involved long and complicated mathematical analysis and took a long time. It yielded, however, some interesting results and ideas which I afterwards found valuable in connection with the theory of the structure of the atom, and also of that of the electric field. The essay was awarded the prize and was published in 1883 by Macmillan under the title *A Treatise on the Motion of Vortex Rings.* [4]

Sir William Thomson's (Lord Kelvin) suggestion that matter might consist of vortex rings of the ether apparently followed his observations of the stability and longevity of smoke rings. J.J. Thomson already had extensive knowledge of the mathematics of vortices and had been familiar with the idea of vortex atoms since his time at Owens: Stewart and Tait had used them in *The Unseen Universe.* As he recalled, the idea appealed to him because of its fundamental nature, raising the possibility that matter, *per se*, was not essential and the whole of physics could be explained using just the ether. His 1882 development of vortex atom theory provided the direct theoretical basis for most of his work until about 1890 (see Chapter 3) and were revived around 1900, along with his ideas of electromagnetic mass, in the electromagnetic view of nature (see Chapter 6).

The bulk of Thomson's *Treatise* is devoted to purely mathematical investigation, but he makes its relevance to vortex atoms explicit in his introductory paragraphs. In the last part of the *Treatise* he develops the theory, qualitatively, into a full-blown explanation of atomic and molecular structure, valency, and chemical combination. An extract from the *Treatise* is reproduced on p. 31. In this work we see, for the first time, Thomson speculating about the possibility of an atom being built up of subatomic entities.

Thomson, educated in the 1870s, grew up with the need for a complex, structured atom to explain spectra. But he was not greatly concerned with this problem; he was far more interested in explaining atomic valency and the periodic properties of the elements. His vortex atom model did this. The model was described in the *Treatise* and again, rather more qualitatively, in 1890 in some manuscript notes on molecular theories. Here he headed it 'A theory of the structure of the molecule which may possibly explain the periodic law.' It is explained in Figures 2.1 and 2.2. He suggested that atoms were

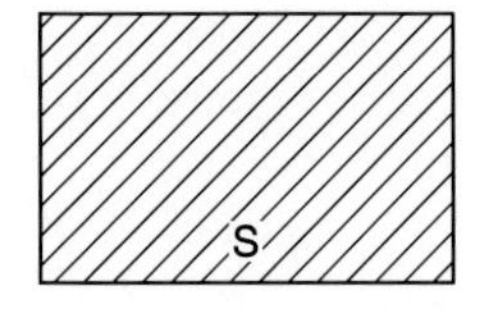

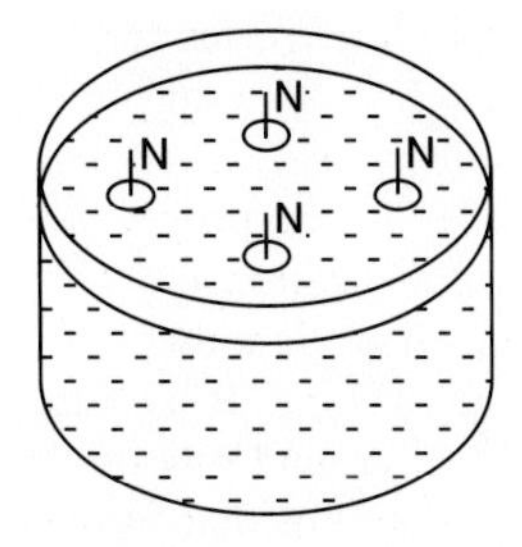

Figure 2.1 Mayer's magnets. Magnetised needles are pushed through cork discs and floated on water. Since they all have their north poles pointing the same way, they repel each other. The south pole of a large magnet is held above the bowl and the needles arrange themselves in a series of concentric rings

composed of groups of interlocking vortex rings. To find the equilibrium position of the rings he drew an analogy to A.M. Mayer's systems of floating magnets. Mayer had found that a collection of magnets floating vertically in a bowl of water with their north poles pointing up, and under the attraction of a large south pole held above the bowl, arranged themselves in a series of concentric rings. The configurations of the rings varied periodically as the number of magnets was increased. Thomson showed that vortex rings arranged themselves in a similar way. He suggested an analogy to the periodic law for the chemical properties of elements:

> If we examine [the figures of equilibrium] we see that as the number of molecules [i.e. vortex filaments] increase there is a tendency for certain peculiarities to occur . . . Thus, if we regard the elements as made up of one substance and increasing atomic weight to indicate a [?new] number of atoms of this primordial element then as the number of atoms is continually increased certain peculiarities in their structure will recur which would probably be accompanied by a recurrence of certain properties. [5]

The confusion of this quotation caused by Thomson's use of the terms 'molecule' and 'atom' is interesting. He used 'molecule' to denote either a molecule or an atom, and 'atom' to denote either an atom or a primordial subatomic particle. This is symptomatic of the general lack of theoretical knowledge at the time about how atoms combined into molecules and about

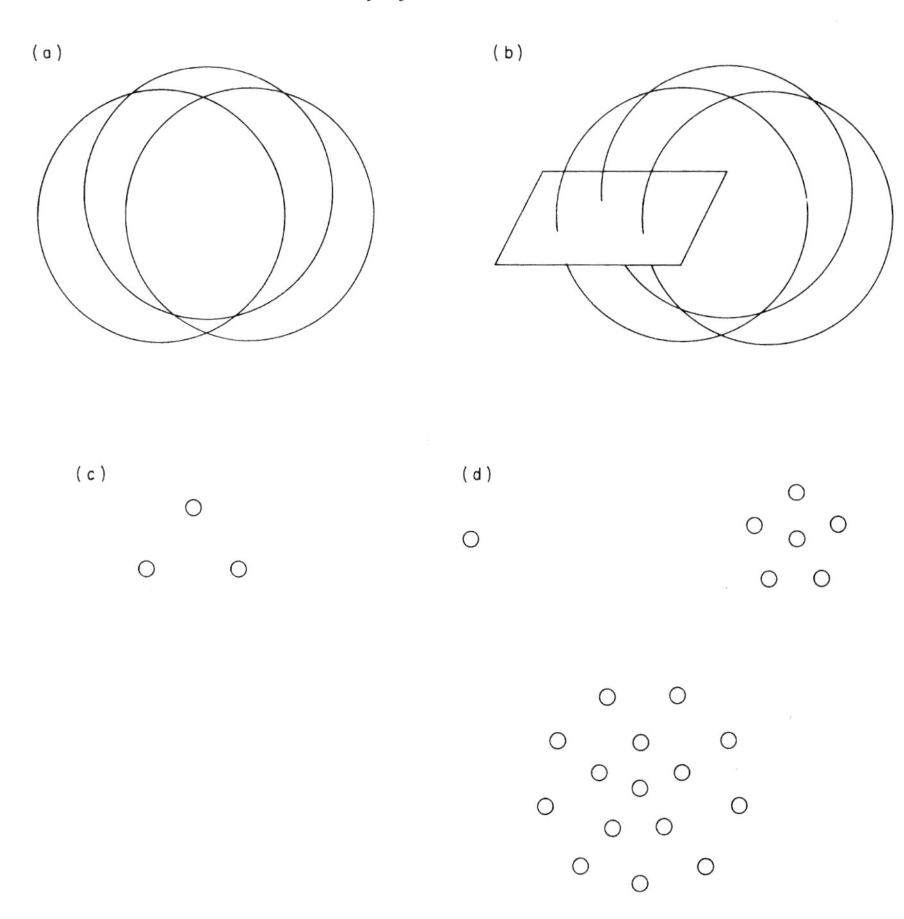

Figure 2.2 Thomson's vortex atom. (a) Consider an atom made up of several vortex rings. (b) Take a plane section through the system. (c) The rings cut the plane in a definite arrangement of points. (d) Mathematical investigation shows that as the number of vortex rings (and hence the number of points) increases, the arrangement of points alters in a periodic manner, just as the arrangement of Mayer's magnets alters: the arrangement of one, six and fifteen rings

the difference between molecular combination and the combination by which primordial elements formed atoms.

The Cavendish Laboratory

Independently of his theoretical researches, Thomson had started some experimental work at the Cavendish, Cambridge's physics laboratory, immediately after taking his degree. Despite his interest in physics he had not previously been there except to pay the occasional visit to Schuster or Poynting, both of whom spent some time in Cambridge. But, from 1880 on, it was to assume an important place in his life.

The Cavendish Laboratory was founded in 1871, one outcome of increasing pressure to bring Cambridge University up to date so that it might continue to provide the administrators of an increasingly technological empire. The Natural Sciences Tripos had been introduced in 1848, one of a series of new Tripos subjects which were beginning to challenge the dominance of the established Mathematics Tripos. It was introduced at a time of intense public and parliamentary scrutiny of Cambridge, paralleling pressures for reform in learned societies, Parliament, and the Civil Service. The Natural Science Tripos was modelled on the Mathematics Tripos: while the Mathematics Tripos taught the 'established' (in William Whewell's terminology) sciences of mechanics and optics, The Natural Sciences Tripos aimed to complement it by teaching the 'progressive' uncertain sciences. Until 1881, students were only allowed to take the Natural Sciences Tripos if they had already graduated in the 'permanent' studies of the Mathematics or Classics Tripos. Despite this provision there was entrenched opposition to the experimental and uncertain nature of the progressive sciences. Isaac Todhunter, a famous mathematical coach and editor of Whewell's collected works, refused an offer by Clerk Maxwell to see an experimental demonstration of conical refraction, 'No. I have been teaching it all my life, and I do not want to have my ideas upset'. [6]

Despite the introduction of experimental sciences, Cambridge's science and, in particular, laboratory provision lagged years behind that in the rest of Britain. Glasgow and London Universities, and Owens College, for example, had far better facilities.

Although a Royal Commission in 1852 stressed the need for practical science teaching at Cambridge, by 1869 nothing had been done. Then a new initiative by the University Senate, recommending the establishment of a University physics laboratory, was blocked by the Colleges who would have had to provide the necessary funding. Only the intervention in 1870 of the Chancellor of the University, William Cavendish 7th Duke of Devonshire, who offered to pay for the Laboratory himself, saved the scheme. The Duke was a wealthy industrialist, a Second Wrangler, and a descendant of Henry Cavendish who had been a pioneer of electrical experiments, had discovered the composition of water, and measured the gravitational constant. A chair of experimental physics, to which Clerk Maxwell was appointed, was established in 1871, and the new laboratory was opened in 1874.

A large part of the difficulty in getting the science laboratories established was that they were, to an extent, pawns in a power struggle between the wealthy colleges and an impoverished university. This struggle has been characterised as between the new class of university professors, advocating recognition of research as a valid academic activity, indeed the academic activity *par excellence*, and the college tutors, trying to preserve a 'gentlemanly' way of life. To the colleges, the laboratories represented the new regime, and they were all too willing to starve them of funds at the slightest provocation. Several colleges preferred to build their own laboratories, over which they had direct control, at far greater expense than it cost them to contribute to a university laboratory.

The Cavendish, alone amongst the university laboratories, had a tradition of financial independence from the University, resulting from its foundation by gift from the Duke of Devonshire. The vitality, which it rapidly developed and became renowned for, may stem from this independence, which enabled it to be the first university laboratory to stress research and professional training, to build up a staff of university lecturers, and to be among the earliest to admit women. In other laboratories, dependent on university funding, the penalties of not toeing the establishment line could be severe. In 1887, for example, James Stuart, Professor of Engineering, was starved of funds and forced to resign his position following his involvement in radical politics. This was partly a manoeuvre designed to put the entire future of the engineering department in jeopardy since the chair was to lapse on Stuart's death or resignation. Only the intervention of the Chancellor, the Duke of Devonshire, saved the chair by pronouncing that it should be a permanent position.

Despite its promising beginnings, Maxwell, the first Cavendish Professor, lacked students to teach, as there were no university regulations requiring even natural sciences students to take a practical exam, and the college dons did not consider the laboratory a fit place for the instruction of gentlemen. Todhunter was known to remark that it was unnecessary for students to see experiments performed, since the results could be vouched for by their teachers, 'all of them men of the highest character, and many of them clergymen of the Church of England'. [7]

Maxwell, therefore, concentrated on building the Cavendish laboratory up as a research institution and here he was lucky. Cambridge had a large supply of mathematicians, hanging around after they graduated in the hope of getting a college fellowship or teaching position, just as did Thomson. Moreover, Trinity College had recently required a research dissertation as part of their fellowship exam and other colleges soon followed suit. These two factors encouraged graduates with leisure and mathematical training to do research for a year or two, and Maxwell exploited them in developing the laboratory. At the same time he pressed for curriculum reform and in 1881, among other changes, practical exams were instituted in the Natural Sciences Tripos.

It was Maxwell's successor, Lord Rayleigh, who benefited from these changes, for Maxwell died in 1879. Lord Rayleigh was appointed to succeed him, as much to raise the perceived social status of the Laboratory as because of his own undoubted status as a mathematician (Senior Wrangler) and physicist. To Rayleigh fell the task of organising undergraduate teaching in the laboratory, which he did along lines which have since become standard in schools and colleges. Suitable standard experiments illustrating aspects of theory and common techniques were devised, and the students were expected to perform these and write up their results in a prescribed manner. This was very different from the way Thomson had been trained under Balfour Stewart,

James Clerk Maxwell

Lord Rayleigh

helping Stewart with his experiments and being encouraged to try out his own ideas, and Thomson always rather lamented the change, while acknowledging that it was probably necessary when dealing with large numbers of students. Rayleigh appointed two demonstrators, Richard Tetley Glazebrook and William Napier Shaw, to institute the practical classes and their experiments were subsequently published in an influential textbook, *Practical Physics*.

Research continued and Rayleigh sought to strengthen the tradition by uniting the workers around a common theme. He chose the establishment of electrical standards, and this is the context within which Thomson performed his own first experiments in the laboratory. These were designed to test a deduction from Maxwell's electromagnetic theory, but he obtained no definite results. Then, at Rayleigh's suggestion, he turned first to some effects produced in the working of induction coils by the electrostatic capacity of the primary and secondary coils, and then to the determination of the ratio of the electrostatic to electromagnetic units of electricity. According to Maxwell's theory the ratio should equal the velocity of light, and much of the promise of Maxwell's theory lay in this potential for unification. However, the equality had not yet been sufficiently established experimentally.

> Rayleigh had already designed some of the apparatus to be used, and had contemplated taking part in the work himself . . . but, as he mentioned to me many years later, 'Thomson rather ran away with it', a natural result of energy, enthusiasm and self-reliance. Work of this kind, which is nothing if not highly accurate, is, however, full of traps and pitfalls, and perusal of the published paper suggests that the author whose experience in experiments was rather limited for so ambitious an undertaking was over sanguine that he had foreseen the possible sources of error, without applying the test of using alternative methods. [8]

In this he recalls Schuster's criticism of Balfour Stewart for trying refined experiments with inadequate apparatus. One remark Thomson makes in the published paper illustrates the lack of provision, even in this comparatively well-endowed laboratory:

> It may be worthy of remark that as many of the pieces of apparatus used were required for the ordinary work of the laboratory, the whole arrangement had to be taken down and put together again between each determination. This must have had the effect of getting rid of a good many accidental errors. [9]

However, despite his comparatively poor showing as an experimentalist, Thomson's promise as a physicist became clear during this period. In 1881 he had applied, unsuccessfully, for the new Professorship of Applied Mathematics at Owens College. The choice evidently lay between him and Arthur Schuster, and Rayleigh was pressed for his opinion:

> Rayleigh saw great merit in both candidates. Roscoe, representing Manchester, wanted him to pronounce in favour of Schuster. 'Why don't you say straight out

that Schuster is the best?' 'I am not sure that he *is* the best', Rayleigh answered. [10]

Now, by 1884, we find Thomson evincing all the traditions that were to become his hallmark; a committed college man, with a reputation for attacking fundamental problems, seeking unification within an ether-based, mechanical, physics; a man with an enormous fertility of theoretical invention and a fairly cavalier approach to experiment.

References

Cambridge University Library holds an important collection of Thomson manuscripts, classmark ADD 7654, referred to here as CUL ADD 7654, followed by the particular manuscript number. Other archives, and a more complete bibliography, may be found in: Falconer, I., Theory and Experiment in J.J. Thomson's work on Gaseous Discharge, PhD Thesis, 1985, University of Bath.

[1] THOMSON, J. J., *Recollections and Reflections*, 1936, London: Bell. Reprinted 1975, New York: Arno, p. 21. Page references given refer to the later edition.
[2] THOMSON, G. P., *J.J. Thomson and the Cavendish Laboratory in his Day*, 1964, London: Nelson, p. 24.
[3] See reference [1], pp. 80–81.
[4] See reference [1], pp. 94–95.
[5] CUL ADD 7654 NB35a.
[6] CROWTHER, J. G., *The Cavendish Laboratory 1874–1974*, 1974, London: Macmillan, p. 9.
[7] Quoted in Macleod, R. and Moseley, R., 'The Naturals' and Victorian Cambridge: Reflections on the Anatomy of an Elite 1851–1914, *Oxford Revue of Education*, **6** (1980), 177–195.
[8] RAYLEIGH, FOURTH LORD, *The Life of Sir J.J. Thomson*, 1942, Cambridge University Press. Reprinted 1969, London: Dawsons, p. 18.
[9] THOMSON, J. J., On the determination of the number of electrostatic units in the electromagnetic unit of electricity. *Philosophical Transactions of the Royal Society*, **174** (1883) 707–721.
[10] STRUTT, R. J., *John William Strutt, Third Baron Rayleigh*, 1924, London: Arnold. Reprinted with additions 1968, Madison: University of Wisconsin Press, pp. 413–414.

THE

LONDON, EDINBURGH, AND DUBLIN

PHILOSOPHICAL MAGAZINE

AND

JOURNAL OF SCIENCE.

[FIFTH SERIES.]

APRIL 1881.

XXXIII. *On the Electric and Magnetic Effects produced by the Motion of Electrified Bodies.* By J. J. THOMSON, *B.A., Fellow of Trinity College, Cambridge**.

§ 1. IN the interesting experiments recently made by Mr. Crookes (Phil. Trans. 1879, parts 1 and 2) and Dr. Goldstein (Phil. Mag. Sept. and Oct. 1880) on "Electric Discharges in High Vacua," particles of matter highly charged with electricity and moving with great velocities form a prominent feature in the phenomena; and a large portion of the investigations consists of experiments on the action of such particles on each other, and their behaviour when under the influence of a magnet. It seems therefore to be of some interest, both as a test of the theory and as a guide to future experiments, to take some theory of electrical action and find what, according to it, is the force existing between two moving electrified bodies, what is the magnetic force produced by such a moving body, and in what way the body is affected by a magnet. The following paper is an attempt to solve these problems, taking as the basis Maxwell's theory that variations in the electric displacement in a dielectric produce effects analogous to those produced by ordinary currents flowing through conductors.

For simplicity of calculation we shall suppose all the moving bodies to be spherical.

* Communicated by the Author.

§ 2. The first case we shall consider is that of a charged sphere moving through an unlimited space filled with a medium of specific inductive capacity K.

The charged sphere will produce an electric displacement throughout the field; and as the sphere moves the magnitude of this displacement at any point will vary. Now, according to Maxwell's theory, a variation in the electric displacement produces the same effect as an electric current; and a field in which electric currents exist is a seat of energy; hence the motion of the charged sphere has developed energy, and consequently the charged sphere must experience a resistance as it moves through the dielectric. But as the theory of the variation of the electric displacement does not take into account any thing corresponding to resistance in conductors, there can be no dissipation of energy through the medium; hence the resistance cannot be analogous to an ordinary frictional resistance, but must correspond to the resistance theoretically experienced by a solid in moving through a perfect fluid. In other words, it must be equivalent to an increase in the mass of the charged moving sphere, which we now proceed to calculate.

Let a be the radius of the moving sphere, e the charge on the sphere, and let us suppose that the sphere is moving parallel to the axis of x with the velocity p; let ξ, η, ζ be the coordinates of the centre of the sphere; let f, g, h be the components of the electric displacement along the axes of x, y, z respectively at a point whose distance from the centre of the sphere is ρ, ρ being greater than a. Then, neglecting the self-induction of the system (since the electromotive forces it produces are small compared with those due to the direct action of the charged sphere), we have

$$f = -\frac{e}{4\pi}\frac{d}{dx}\frac{1}{\rho},$$

$$g = -\frac{e}{4\pi}\frac{d}{dy}\frac{1}{\rho},$$

$$h = -\frac{e}{4\pi}\frac{d}{dz}\frac{1}{\rho};$$

therefore

$$\frac{df}{dt} = -\frac{ep}{4\pi}\frac{d^2}{dx\,d\xi}\frac{1}{\rho},$$

$$\frac{dg}{dt} = -\frac{ep}{4\pi}\frac{d^2}{d\xi\,dy}\frac{1}{\rho},$$

$$\frac{dh}{dt} = -\frac{e}{4\pi}\frac{d^2}{d\xi\,dz}\frac{1}{\rho};$$

hence

$$\left.\begin{aligned}\frac{df}{dt}&=\frac{ep}{4\pi}\frac{d^2}{dx^2}\frac{1}{\rho},\\ \frac{dg}{dt}&=\frac{ep}{4\pi}\frac{d^2}{dx\,dy}\frac{1}{\rho},\\ \frac{dh}{dt}&=\frac{ep}{4\pi}\frac{d^2}{dx\,dz}\frac{1}{\rho}.\end{aligned}\right\}\quad\ldots\ldots\quad(1)$$

Using Maxwell's notation, let F, G, H be the components of the vector-potential at any point; then, by 'Electricity and Magnetism,' § 616,

$$F=\mu\iiint\frac{u}{\rho'}\,dx\,dy\,dz,$$

$$G=\mu\iiint\frac{v}{\rho'}\,dx\,dy\,dz,$$

$$H=\mu\iiint\frac{w}{\rho'}\,dx\,dy\,dz,$$

where u, v, w are the components of the electric current through the element $dx\,dy\,dz$, and ρ' is the distance of that element from the point at which the values of F, G, H are required, μ is the coefficient of magnetic permeability. In the case under consideration,

$$F=\mu\iiint\frac{\frac{df}{dt}}{\rho'}\,dx\,dy\,dz\,;$$

substituting for $\frac{df}{dt}$ its value from equation (1), we get

$$F=\frac{\mu ep}{4\pi}\iiint\frac{1}{\rho'}\frac{d^2}{dx^2}\frac{1}{\rho}\,dx\,dy\,dz,$$

with similar expressions for G and H.

Let us proceed to calculate the value of F at a point P.

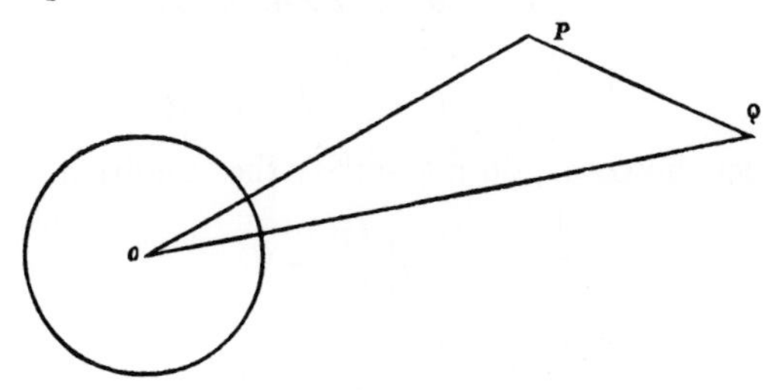

Let O be the centre of the sphere; then OQ$=\rho$, PQ$=\rho'$, OP$=$R,

$$F = \frac{\mu e p}{4\pi} \iiint \frac{1}{PQ} \frac{d^2}{dx^2} \frac{1}{\rho} \, dx \, dy \, dz.$$

Now $\frac{d^2}{dx^2} \frac{1}{\rho} = \frac{Y_2}{\rho^3}$, where Y_2 is a surface harmonic of the second order. And when $\rho > R$,

$$\frac{1}{PQ} = \frac{1}{\rho} + \frac{R}{\rho^2} Q_1 + \frac{R^2}{\rho^3} Q_2 + \ldots ;$$

and when $\rho < R$,

$$\frac{1}{PQ} = \frac{1}{R} + \frac{\rho}{R^2} Q_1 + \frac{\rho^2}{R^3} Q_2 + \ldots ;$$

where Q_1, Q_2, &c. are zonal harmonics of the first and second orders respectively referred to OP as axis.

Let Y'_2 denote the value of Y_2 along OP. Then, since $\int Y_n Q_m \, ds$, integrated over a sphere of unit radius, is zero when n and m are different, and $\frac{4\pi}{2n+1} Y'_n$ when $n = m$, Y'_n being the value of Y_n at the pole of Q_n, and since there is no electric displacement within the sphere,

$$F = \frac{\mu e p}{4\pi} \times \frac{4\pi Y'_2}{5} \left\{ \int_R^\infty \frac{R^2}{\rho^4} \, d\rho + \int_a^R \frac{\rho \, d\rho}{R^3} \right\}$$

$$= \frac{\mu e p}{5} Y'_2 \left(\frac{5}{6R} - \frac{a^2}{2R^3} \right),$$

or, as it is more convenient to write it,

$$= \frac{\mu e p}{5} \left(\frac{5R^2}{6} - \frac{a^2}{2} \right) \frac{d^2}{dx^2} \frac{1}{R}.$$

By symmetry, the corresponding values of G and H are

$$G = \frac{\mu e p}{5} \left(\frac{5R^2}{6} - \frac{a^2}{2} \right) \frac{d^2}{dx \, dy} \frac{1}{R},$$

$$H = \frac{\mu e p}{5} \left(\frac{5R^2}{6} - \frac{a^2}{2} \right) \frac{d^2}{dx \, dz} \frac{1}{R}.$$

These values, however, do not satisfy the condition

$$\frac{dF}{dx} + \frac{dG}{dy} + \frac{dH}{dz} = 0.$$

If, however, we add to F the term $\frac{2\mu e p}{3R}$, this condition will be satisfied; while, since the term satisfies Laplace's equation, the other conditions will not be affected: thus we have finally for points outside the sphere,

$$\left.\begin{aligned} F &= \frac{\mu ep}{5}\left(\frac{5R^2}{6}-\frac{a^2}{2}\right)\frac{d^2}{dx^2}\frac{1}{R}+\frac{2\mu ep}{3R}, \\ G &= \frac{\mu ep}{5}\left(\frac{5R^2}{6}-\frac{a^2}{2}\right)\frac{d^2}{dx\,dy}\frac{1}{R}, \\ H &= \frac{\mu ep}{5}\left(\frac{5R^2}{6}-\frac{a^2}{2}\right)\frac{d^2}{dx\,dz}\frac{1}{R}. \end{aligned}\right\} \quad . \quad . \quad (2)$$

Now, by 'Electricity and Magnetism,' § 634, T the kinetic energy

$$=\tfrac{1}{2}\iiint(Fu+Gv+Hw)dx\,dy\,dz,$$

in our case,

$$=\tfrac{1}{2}\iiint\left(F\frac{df}{dt}+G\frac{dg}{dt}+H\frac{dh}{dt}\right)dx\,dy\,dz.$$

Now

$$\tfrac{1}{2}\iiint F\frac{df}{dt}\,dx\,dy\,dz,$$

substituting for F and $\frac{df}{dt}$,

$$=\tfrac{1}{2}\frac{\mu e^2p^2}{20\pi}\iiint\left(\frac{5r^2}{6}-\frac{a^2}{2}\right)\left(\frac{d^2}{dx^2}\frac{1}{r}\right)^2dx\,dy\,dz,$$

since the term

$$\frac{\mu e^2p^2}{12\,.\,\pi}\iiint\frac{1}{r}\frac{d^2}{dx^2}\frac{1}{r}\,dx\,dy\,dz$$

evidently vanishes.

Transforming to polars and taking the axis of x as the initial line, the above integral

$$=\frac{\mu e^2p^2}{40\pi}\int_0^{2\pi}\int_0^{\pi}\int_a^{\infty}\left(\frac{5r^2}{6}-\frac{a^2}{2}\right)\frac{(3\cos^2\theta-1)^2}{r^4}\sin\theta\,dr\,d\theta\,d\phi$$

$$=\frac{4\mu e^2p^2}{75a},$$

$$\tfrac{1}{2}\iiint G\frac{dg}{dt}\,dx\,dy\,dz=\frac{\mu e^2p^2}{40\pi}\iiint\left(\frac{5}{6}r^2-\frac{a^2}{2}\right)\left(\frac{d^2}{dx\,dy}\frac{1}{r}\right)^2dx\,dy\,dz.$$

By transforming to polars, as before, we may show that this

$$=\frac{\mu e^2p^2}{25a}.$$

Similarly,

$$\tfrac{1}{2}\iiint H\frac{dh}{dt}\,dx\,dy\,dz=\frac{\mu e^2p^2}{25a};$$

∴ T, the kinetic energy due to the electrification

$$= \frac{1}{2}\iiint\left(\mathrm{F}\frac{df}{dt} + \mathrm{G}\frac{dg}{dt} + \mathrm{H}\frac{dh}{dt}\right) dx\,dy\,dz$$

$$= \frac{2\mu e^2 p^2}{15a}.$$

Hence, if m be the mass of the sphere, the whole kinetic energy

$$= \left(\frac{m}{2} + \frac{2}{15}\frac{\mu e^2}{a}\right)p^2; \quad . \quad . \quad . \quad . \quad . \quad (3)$$

or the effect of the electrification is the same as if the mass of the sphere were increased by $\frac{4}{15}\frac{\mu e^2}{a}$, or, if V be the potential of the sphere, by $\frac{4}{15}\mu \mathrm{K}^2 \mathrm{V}^2 a$.

To form some idea of what the increase of mass could amount to in the most favourable case, let us suppose the earth electrified to the highest potential possible without discharge, and calculate the consequent increase in mass. According to Dr. Macfarlane's experiments, published in the Philosophical Magazine for December 1880, the electric force in air at ordinary temperatures and pressures must not exceed 3×10^{12} (electromagnetic system of units). The electric force just outside the sphere is V/a; hence the greatest possible value of V is $3 \times 10^{12} a$, where a is the radius of the earth. Putting this value for V, $\mu = 1$, $\mathrm{K} = \frac{1}{9 \cdot 10^{20}}$, $a = 6 \cdot 4 \times 10^8$, we get for the corresponding value of the increase of mass 7×10^8 grms., or about 650 tons, a mass which is quite insignificant when compared with the mass of the earth.

For spheres of different sizes, the greatest increase in mass varies as the cube of the radius; hence the ratio of this increase to the whole mass of the sphere is constant for all spheres of the same material; for spheres of different materials the ratio varies inversely as the density of the material.

If the body moves so that its velocities parallel to the axes of x, y, z respectively are p, q, r, then it is evident that the effect of the electrification will be equivalent to an increase of $\frac{4}{15}\mu \mathrm{K}^2 \mathrm{V}^2 a(p^2 + q^2 + r^3)$ in the mass of the sphere.

ON THE MOTION OF VORTEX RINGS.

§ 1. THE theory that the properties of bodies may be explained by supposing matter to be collections of vortex lines in a perfect fluid filling the universe has made the subject of vortex motion at present the most interesting and important branch of Hydrodynamics. This theory, which was first started by Sir William Thomson, as a consequence of the results obtained by Helmholtz in his epoch-making paper "Ueber Integrale der hydrodynamischen Gleichungen welche den Wirbelbewegungen entsprechen" has *à priori* very strong recommendations in its favour. For the vortex ring obviously possesses many of the qualities essential to a molecule that has to be the basis of a dynamical theory of gases. It is indestructible and indivisible; the strength of the vortex ring and the volume of liquid composing it remain for ever unaltered; and if any vortex ring be knotted, or if two vortex rings be linked together in any way, they will retain for ever the same kind of be-knottedness or linking. These properties seem to furnish us with good materials for explaining the permanent properties of the molecule. Again, the vortex ring, when free from the influence of other vortices, moves rapidly forward in a straight line; it can possess, in virtue of its motion of translation, kinetic energy; it can also vibrate about its circular form, and in this way possess internal energy, and thus it affords us promising materials for explaining the phenomena of heat and radiation.

This theory cannot be said to explain what matter is, since it postulates the existence of a fluid possessing inertia; but it proposes to explain by means of the laws of Hydrodynamics all the properties of bodies as consequences of the motion of this fluid. It is thus evidently of a very much more fundamental character than any theory hitherto started; it does not, for example, like the ordinary kinetic theory of gases, assume that the atoms attract each other with a force which varies as that power of the distance which is most convenient, nor can it hope to explain any property of bodies by giving the same property to the atom. Since this theory is the only one that attempts to give any account of the mechanism of the intermolecular forces, it enables us to form much the clearest mental representation of what goes on when one atom influences another. Though the theory is not sufficiently developed for us to say whether or not it succeeds in explaining all the properties of bodies, yet, since it gives to the subject of vortex motion the greater part of the interest it possesses, I shall not scruple to examine the consequences according to this theory of any results I may obtain.

The present essay is divided into four parts: the first part, which is a necessary preliminary to the others, treats of some general propositions in vortex motion and considers at some length the theory of the single vortex ring; the second part treats of the mutual action of two vortex rings which never approach closer than a large multiple of the diameter of either, it also treats of the effect of a solid body immersed in the fluid on a vortex ring passing near it; the third part treats of knotted and linked vortices; and the fourth part contains a sketch of a vortex theory of chemical combination, and the application of the results obtaining in the preceding parts to the vortex ring theory of gases.

It will be seen that the work is almost entirely kinematical; we start with the fact that the vortex ring always consists of the same particles of fluid (the proof of which, however, requires dynamical considerations), and we find that the rest of the work is kinematical. This is further evidence that the vortex theory of matter is of a much more fundamental character than the ordinary solid particle theory, since the mutual action of two vortex rings can be found by kinematical principles, whilst the "clash of atoms" in the ordinary theory introduces us to forces which themselves demand a theory to explain them.

have scarcely been taken sufficiently into account to make the experiment the crucial test of a theory; and it is probable that the theory of the diffusion and viscosity of gases worked out from the laws of action of two vortex rings on each other given in Part II of this essay would lead to results which would decide more easily and more clearly between the two theories.

The preceding reasoning holds only for a monatomic gas which can only increase its energy by increasing the mean radius of its vortex atoms; if however the gas be diatomic the energy will be increased if the shortest distance between the central lines of the vortex cores of the two atoms be diminished, and if the radius of the vortex atom is unaltered the velocity of translation of the molecule will be increased as well as the energy; thus for a diatomic molecule we cannot say that an increase in the energy or a rise in the temperature of the gas would necessarily be accompanied by a diminution in the mean velocity of its molecules.

§ 58. We shall now go on to apply some of the foregoing results to the case of chemical combination; in the following remarks we must be understood to refer only to bodies in the gaseous state. When two vortex rings of equal strength, with (as we shall suppose for simplicity) their planes approximately parallel to each other and approximately perpendicular to the line joining their centres, are moving in the same direction, and the circumstances are such that the hinder ring overtakes the one in front, then if, when it overtakes it, the shortest distance between the circular lines of vortex core of the rings be small compared with the radius of either ring, the rings will not separate, the shortest distance between their central lines of vortex core will remain approximately constant, and these central lines of vortex core will rotate round another circle midway between them, while this circle moves forward with a velocity of translation which is small compared with the linear velocity of the vortex rings round it. We may suppose that the union or pairing in this way of two vortex rings of different kinds is what takes place when two elements of which these vortex rings are atoms combine chemically; while, if the vortex rings are of the same kind, this process is what occurs when the atoms combine to form molecules. If two vortex rings paired in the way we have described are subjected to any disturbing influence, such as the action due to other vortex rings in their neighbourhood, their radii will be changed by different amounts; thus their velocities of translation will become different, and they will separate. We are thus led to take the view of chemical combination put forward by Clausius and Williamson, according to which the molecules of a compound gas are supposed not to always consist of the same atoms of the elementary gases, but that these atoms are continually changing partners. In order, however, that the compound gas should be something more than a mechanical mixture of the elementary gases of which it is composed, it is evidently necessary that the mean time during which an atom is paired with another of a different kind, which we shall call the paired time, should be large compared with the time during which it is alone and free from other atoms, which we shall call the free time. If we suppose that the gas is subjected to any disturbance, then this will have the effect of breaking up the molecules of the compound gas sooner than would otherwise be the case. It will thus diminish the ratio of the paired to the free time; and if the disturbance be great enough, the value of this ratio will be so much reduced that the substance will no longer exhibit the properties of a chemical compound, but those of its constituent elements: we should thus have the phenomenon of dissociation or decomposition.

We know that when two elements combine a large amount of heat is in many cases given out. We have proved in § (56) that for two vortex rings in the position of the vortex atoms of a molecule of a chemical compound $\Sigma\left(f\frac{d\mathfrak{P}}{dt}+g\frac{d\mathfrak{Q}}{dt}+h\frac{d\mathfrak{R}}{dt}\right)$ is positive; when the vortex rings are separated by a distance very great compared with the radius of either this quantity vanishes: thus we see from equation (9) that $\Sigma\mathfrak{J}V'$ is increased by the combination of the atoms so that this would explain the evolution of a certain amount of heat. I do not think however that this cause would account for the enormous quantities of heat generated in some cases of chemical combination, for even these large as they are seem only to be the differences between quantities much greater than themselves. Thus for example the heat given out when hydrogen and chlorine combine to form hydrochloric acid is the difference between the heat given out when the atoms of hydrogen combine with the atoms of chlorine to form hydrochloric acid and the heat required to split up the hydrogen and chlorine molecules into their atoms. The determinations by Prof. E. Wiedemann of the heat given out when hydrogen atoms combine to form molecules, and by Prof. Thomson of the same quantity for carbon atoms seem to shew that these quantities are much greater than the quantities of heat given out in ordinary chemical combinations, and thus that these latter quantities are the differences of quantities much greater than themselves.

Whatever be the reason, the pairing of two atoms, whether of the same or different kinds, is attended by a large increase in the translatory energy. The vortex atoms however do not remain continually paired, and two atoms will only contribute to the increase in the translatory energy whilst they are paired and not when they are free, thus the whole increase in the translatory energy of a large number of molecules will depend not only on the amount of the increase contributed by any two atoms when they pair, but also on the time they remain together, and will thus depend on the ratio of the paired to the free times for the substance. The ratio of the paired to the free time plays also a very important part in determining whether chemical combination shall take place or not, and when it does take place the proportion between the amounts of the various compounds formed when more compounds than one are possible. It is clear too that the value of this ratio for the atoms of an elementary gas will have a very great effect on the chemical properties of the gas : thus if the ratio of the free to the paired times for the atoms of the gas be very small the gas will not enter readily into combination with other gases, for it will only do so to any great extent when the ratio of the free to the paired time for the compound is less than for the atoms in the molecule of the elementary gas, but if the latter be very small there is less likelihood of the ratio for the compound gas being less ; thus we should expect that this ratio would be very small for the atoms of a gas like nitrogen which does not combine readily with other gases. The value of the ratio would afford a very convenient measure for the affinity of the constituents of a compound for each other. It is also conceivable that this ratio might affect the physical properties of a gas, and in a paper in the *Philosophical Magazine* for June 1883 I suggested that differences in the value of this ratio might account for the differences in the dielectric strengths of gases.

Two vortex rings will not remain long together unless the shortest distance between the central lines of their vortex cores is small compared with the radius of either of the rings; now as the vortex rings approach each other they alter in size, the one in front expands and the one in the rear contracts. If the rings are to remain together their radii must become nearly equal as they approach each other and their planes become nearly coincident: it is evident however that for this to happen the radii of the rings before they pair must lie within certain limits. The energy of the gas however, and therefore the temperature depend upon the mean radius of the vortex rings which form the atoms of the gas, and conversely the mean radius of the vortex atoms is a function of the temperature, and if the mean radius is between certain limits the temperature must also be between limits, thus unless the temperature is between certain limits the atoms would not remain long together after they paired and so chemical combination would not take place; this reasoning would indicate that chemical combination could only occur between certain limits of temperature, and this seems to be the case in at any rate a great many cases of chemical combination.

The following reasoning will explain how it is that the compound after it is formed can exist at temperatures at which the elements of which it is composed could not combine. When the elements have once combined the molecules of the compound will settle down so that the radii of their vortex atoms will be distributed according to a definite law, and a large proportion of the vortex atoms will have their radii between comparatively narrow limits, just as in the ordinary theory of gases Maxwell's law gives the distribution of velocity. Now suppose that a molecule of a compound of the elements A and B is subjected to any disturbance tending to change the radii of the atoms; though the difference in the changes in the radii may be sufficient to cause the atoms to separate, yet since the atoms were close together when they were disturbed the difference in the changes must be small, and since the motion is reversible the atom A would only have to suffer a slight change to be able to combine again with a vortex ring like B, or it could combine at once with a vortex ring differing only slightly in radius from B; thus A will have plenty of chances of recombination with the B atoms and will be in a totally different position with regard to them from that in which it would have been if it had not previously been in combination with a B atom.

Let us now suppose that two vortex rings of approximately equal radius but of different strengths come close together in such a way that their planes are approximately parallel and perpendicular to the line joining their centres, then we can see, as in the analogous case of linked vortices, that the motion will be of the following kind. Let m and m' be the strengths of the two vortices, a the mean radius of either, and d the shortest distance between their central lines, e and e' the radii of the cross sections of the two vortex rings. If we imagine a circle between the central lines of the two vortex rings dividing the distance between the vortices inversely as the strengths of the vortices, the two vortex rings will rotate round this circle with an angular velocity $m+m'/d\pi^2$ remaining at an approximately constant distance d apart, while the circle itself will move with a comparatively slow motion of translation perpendicular to its own plane.

The mean velocity of the vortex of strength m

$$=\frac{m}{2\pi a}\left(\log\frac{8a}{e}-1\right)+\frac{m'}{2\pi a}\left(\log\frac{8a}{d}-1\right);$$

the mean velocity of the vortex of strength m'

$$=\frac{m'}{2\pi a}\left(\log\frac{8a}{e'}-1\right)+\frac{m}{2\pi a}\left(\log\frac{8a}{d}-1\right).$$

Now if the two vortices are to remain together, their mean velocities must be equal; therefore

$$\frac{m}{2\pi a}\log\frac{8a}{e}+\frac{m'}{2\pi a}\log\frac{8a}{d}=\frac{m'}{2\pi a}\log\frac{8a}{e'}+\frac{m}{2\pi a}\log\frac{8a}{d}.$$

Now suppose a and d become, through some external influence, $a+\delta a$ and $d+\delta d$, then the change in the mean velocity of the vortex of strength m is, if V be the original mean velocity,

$$\left(-\frac{V}{a}+\frac{3m+2m'}{4\pi a^2}\right)\delta a-\frac{m}{2\pi a}\cdot\frac{\delta d}{d};$$

and if we interchange m and m' in this formula, we shall get the change in the mean velocity of the vortex ring whose strength is m'. Now if the two vortex rings are to remain together for a time long compared with the mean interval between two collisions, in spite of all the vicissitudes they will meet with when moving about in an enclosure containing a great number of moving molecules, the mean velocities of the two vortex rings must always remain equal; thus the changes in the mean velocities of the rings must be equal for all values of δa and δd, so that the coefficients of δa and δd must be equal in the two expressions for the changes in the mean velocities; for this to be the case we see that m must equal m'. Hence, we conclude that if two vortex rings are to remain for long together when subject to disturbing influences, they must be of equal strength. We can extend this result to the case when we have more than two vortices close together; however many vortices there are, if they are to remain together for any considerable time they must be of equal strength.

§ 59. We shall often have occasion to speak of vortex rings arranged in the way discussed in § 43, *i.e.* so that those portions of the central lines of vortex core of the several vortex rings which are closest together are always approximately parallel, and so that a plane perpendicular to their central lines at any point cuts them in the angular points of a regular polygon. We proved in Part III that if the vortices are of equal strengths, and not more than six in number, they will be in stable steady motion; it is not necessary for the truth of this proposition that each vortex ring should be single; the proposition will be true if the vortex rings are composite, provided the distances between their components are small compared with the sides of the polygon, at the angular points of which the vortices are situated, and that the sum of the strengths of the components is the same as the strength of the single vortex ring, which they are supposed to replace. We shall speak of the systems of vortices placed at the angular points of the polygon as the primaries, and the component vortex rings of these primaries as the secondaries of the system; and when we speak of a system consisting of three, four, five, or six primaries, we shall suppose, unless we expressly state the contrary, that they are arranged in the way just described.

We may imagine the way in which these vortex rings are linked through each other by supposing that we take a cylindrical rod and describe on its surface a screw with n threads; let us first suppose that there are an exact number of turns of each thread on the rod, bend the rod into a circle and join the ends, then each of the n threads of the screw will represent the central line of the vortex core of one of the n equal linked vortices; next suppose that the threads make m/n turns in the length of the rod where m is an integer not divisible by n, then if we bend the rod as before and join the ends, the threads of the screw will form an endless thread with n loops, and for the present purpose the properties of a vortex ring whose core is of this kind will be similar to those of one where the n threads are distinct, so that we may suppose the core of the vorticity which forms the atom to be arranged, in either of these ways, and we shall speak of it as an atom with n links; thus the links may be separate or run one into the other forming an endless chain.

Now let us suppose that the atoms of the different chemical elements are made up of vortex rings all of the same strength, but that some of these elements consist of only one of these rings, others of two of the rings linked together, or else of a continuous curve with two loops, others of three, and so on; but our investigation at the end of Part III shews that no element can consist of more than six of these rings if they are arranged in the symmetrical way there described.

Then if any of these atoms combine so as to form a permanent combination, the strengths of all the primaries in the system formed by the combination must be equal. Thus an atom of an element may combine with another atom of the same kind to form a molecule of the substance consisting of two atoms. Again, three of these atoms may combine and form a system consisting of three primary elements, but the chance of their doing this is small compared with the chance of two pairing, so that the number of systems of this kind will be small compared with the number of the systems consisting of only two atoms. We might have systems consisting of four atoms, but the number would be small compared with the number of systems that consist of three atoms, and so on. We could not have a system consisting of more than six primaries if arranged in the way supposed, but though this seems the most natural way of arranging the atoms, we must not be understood to assert that this is the only way, and in special cases the atoms may be arranged differently, and then we might have systems consisting of more than six primaries. Now, suppose that an atom of one element is to combine with an atom of another. Suppose to fix our ideas, that the atom consisting of two vortex rings linked together is to combine with an atom consisting of one vortex ring, then since for stability of connection, the strength of all the primaries which form the components of the compound system must be equal; the atom consisting of two links must unite with molecules containing two atoms of the one with one link. If the atoms are made to combine directly, the chance that they form the simplest combination is almost infinitely greater than the chance of any more complex combination, so that the number of the simplest compound systems will be almost infinitely greater than the number of any more complex compound system. Thus the compound formed will be the simplest combination, consisting of one of the atoms, which consist of two vortex rings linked together, with two of the atoms consisting of only one vortex ring. Similarly, if an atom consisting of three vortex rings linked together were to combine directly with atoms consisting of only one vortex ring, the compound formed would consist of one of the three linked atoms with three of the others, and so on for the combination of atoms formed by any number of vortex rings linked together. This suggests, that the atoms of the elements called by the chemists monads, dyads, triads, tetrads, and so on, consist of one, two, three, four, &c., vortex rings linked together, for then we should know that a dyad could not combine with less than two atoms of a monad to form a stable compound, a triad with less than three, and so on, which is just the definition of the terms monad, dyad, &c.

Thus each vortex ring in the atom would correspond to a unit of affinity in the chemical theory of quantivalence. If we regard the vortex rings in those atoms consisting of more vortex rings than one as linked together in the most symmetrical way, then no element could have an atom consisting of more than six vortex rings at the most, so that no single atom would be capable of uniting with more than six atoms of another element so as to form a stable compound. This agrees with chemical facts, as Lothar Meyer in his *Modernen Theorien der Chemie*, 4th Edition, p. 196, states that no compound consisting of more than six atoms of one element combined with only one of another is known to exist in the gaseous state, and that a gaseous compound of tungsten, consisting of six atoms of chlorine united to one of tungsten does exist.

Though in direct combination, the simplest compound is the one that would naturally be formed; yet other compounds are possible, and under other circumstances might be formed; thus one atom of a dyad might unite not only with two atoms of a monad, but also with four atoms of a monad, the four atoms of the monad splitting up into two groups of two each; thus the one atom of the dyad and the two groups of two monads would form three primaries, which when arranged in the way described above, would be in stable steady motion; again, we might have two atoms of the dyad and two of the monad forming again a system with three primaries, or one atom of the dyad might unit with one atom of another dyad, forming a system with two primaries, or with two atoms of another dyad forming a system with three primaries, and so on. These remarks may be illustrated by means of the following gaseous compounds of sulphur and mercury.

Thus we have the compounds:

$$H_2S,\ SO_2,\ SO_3,$$
$$Hg.Cl_2,\ Hg_2Cl_2.$$

In fact, all that is necessary for the existence of any compound from this point of view is, that its constituents should be capable of division into primaries of equal strength, and if these are to be arranged in the simplest and most symmetrical way, that there should not be more than six of them.

§ 60. Looking at chemical combinations from this point of view, we should expect to find that such compounds as hydrochloric acid, where one atom of hydrogen has only to meet with one atom of chlorine; or water where an atom of oxygen has only to meet with two atoms or a molecule of hydrogen, would be much more easily and quickly formed by direct combination, than a compound such as ammonia gas, to form which, an atom of nitrogen has to find itself close to three atoms of hydrogen at once; and it is, I believe, the case in direct combination, that simple compounds are formed more quickly than complex ones.

We shall call the ratio of the number of links in the atom of an element to the number in the atom of hydrogen, the valency of the element. To determine this quantity with any degree of certainty, we require to know the accurate composition of a large number of the gaseous compounds of the element; thus only those compounds whose vapour-density is known afford us any assistance, as it would make a great difference, for example, in the valency of nitrogen, if the molecule of ammonia could be represented by the formula N_2H_6 instead of NH_3, and differences of this kind can only be determined by vapour-density determinations; so that in the following discussion of the valency of the elements, too much importance must not be attached to the result for any element, when the vapour densities of only a few of its compounds have been determined. The determination of a single vapour-density will enable us to assign a superior limit to the valency of the elements in the compound, but it may require a great many vapour-density determinations to enable us to assign a lower limit to the valency of the same element.

The compounds HCl, HI, HBr, HF, Tl Cl. shew that the atoms of chlorine, iodine, fluorine, and thallium have the same number of links as the atom of hydrogen, or that the valency of each of these elements is unity. From the compound H_2O we infer that the atom of oxygen consists of twice as many links as the atom of hydrogen, though as far as this compound goes there is nothing to shew that the atom of oxygen does not consist of the same number of links as the atom of hydrogen, in this case however we should have to look upon the molecule of water as a system with the three primaries H – H – O; it is however preferable to take the simpler view that the water molecule is a system with the two primaries HH – O, and suppose that the valency of oxygen is two: the composition of all the compounds of oxygen may be explained on this supposition, and there are other considerations which lead us to endeavour to reduce the number of primaries in the molecule of a compound to as few as possible. Regarding oxygen then as a dyad, the molecule of hydrogen peroxide consists of the three primaries H_2 – O – O.

The compounds H_2S, H_2Se, Pb Cl_2, Cd Br_2, Te H_2, indicate that the atoms of sulphur, selenium, tellurium, lead and cadmium have twice as many links as the atom of hydrogen. The compound CO shews that the atom of carbon has the same number of links as the atom of oxygen, or twice as many as the atom of hydrogen; the molecules of carbonic acid and marsh gas have each three primaries represented by C – O – O and C – H_2 – H_2 respectively. Carbon is usually regarded as a tetrad, and we should therefore have expected its atom to have four times as many links as the atom of hydrogen; the compound CO shews however that if the view we have taken be correct, the carbon atom must have only twice as many links as the hydrogen atom: this view is supported by the composition of acetylene C_2H_2; if the valency of carbon atom be two, the molecule may be divided into the three primaries C – C – H_2, but if the valency of carbon were four, the molecule of acetylene could not be divided into primaries of equal strength, so that according to our view, its constitution is impossible on this supposition.

The sulphur compounds afford good examples of molecules containing various numbers of primaries, thus we have H_2S with two primaries H_2 – S; SO_2 with three primaries S – O – O and SO_3 with four primaries S – O – O – O.

It is difficult to determine from the composition of the mercury compounds as given in the chemical text-books whether the atom of mercury has the same number of links as the atom of hydrogen or twice that number; according to Lothar Meyer the composition of calomel is Hg Cl, in most of the other text-books it is given as Hg_2 Cl_2; if Lothar Meyer's supposition be correct then the mercury atom has as many links as the hydrogen atom and the molecule of calomel consists of the two primaries Hg – Cl while the molecule of corrosive sublimate consists of the three primaries Hg – Cl – Cl; if however the composition of calomel is $Hg_2 Cl_2$ then the mercury atom probably has twice as many links as the hydrogen atom and the molecule of calomel consists of the three primaries Hg – Hg – Cl_2 while the molecule of corrosive sublimate consists of the two primaries Hg – Cl_2.

The following reasons lead us to suppose that the atom of phosphorus has the same number of links as the atom of hydrogen; the composition of phosphoretted hydrogen PH_3 shews that the atom of phosphorus must either have the same number of links as the hydrogen atom in which case the molecule consists of four primaries, or it must have three times as many in which case the molecule of phosphoretted hydrogen will have two primaries; the compound PH_5 however shews that the phosphorus atom has either the same number of links as the hydrogen atom or five times as many; hence we see that the phosphorus atom must have the same number of links as the hydrogen atom. The resemblance between the properties of arsenic and phosphorus would lead us to conclude that the atom of arsenic had the same number of links as the atom of hydrogen, and the constitution of its compounds could be explained on this supposition; there is nothing to shew from its simpler inorganic compounds that the arsenic atom has not three times as many links as the hydrogen atom; the composition of the chloride of cacodyl As Cl $C_2 H_6$ shews however that this is not the case and the atom of arsenic like that of phosphorus must have the same number of links as the hydrogen atom.

The compounds of nitrogen present great difficulties when considered from this point of view; the composition of ammonia NH_3 requires us to suppose either that the nitrogen atom has three times as many links as the hydrogen atom, in which case the molecule of ammonia would consist of the two primaries N – H_3, or that the nitrogen atom has the same number of links as the hydrogen atom and then the molecule of ammonia would consist of the four primaries N – H – H – H; the composition of nitric oxide NO however compels us to suppose that the atom of nitrogen has the same number of links as the atom of oxygen or twice as many as the atom of hydrogen, and these suppositions are inconsistent. It is however conceivable that an atom might go through a process that would cause it to act like one with twice as many links. To illustrate this take a single circular ring and pull the opposite sides so that they cross at the centre of the ring, forming a figure of eight, then bend one half of the figure of eight over the other half, the continuous ring will now form two circles whose planes are nearly coincident. If the circular ring represented a line of vortex core the duplicated ring would behave like one with twice as many links as the original ring. Thus if we look upon the atom of nitrogen as consisting of the same number of links as the atom of hydrogen we can explain the constitution of the compounds NH_3, N_2O, N_2O_3, C_2N_2, HCN, CNH_5 &c., but in the compounds NO, NO_2 we should have to suppose that the atom was duplicated in the manner described above.

The following table shews the valency of those elements which form gaseous compound of known vapour density, though as we said before when we know the vapour density of only a few of the compounds of an element the value given in the table must not be looked on as anything more than an upper limit to the value of the valency of the element.

Univalent Elements.

Arsenic.	Mercury ?
Bromine.	Nitrogen.
Chlorine.	Phosphorus.
Fluorine.	Potassium.
Hydrogen.	Rubidium.
Iodine.	Thallium.

Divalent Elements.

Cadmium.	Mercury ?
Carbon.	Oxygen.
Chromium.	Selenium.
Copper.	Sulphur.
Lead.	Tellurium.
Manganese.	Zinc.

Trivalent Elements.

Aluminium.	Bismuth.
Antimony.	Boron.

Indium.

Quadrivalent Elements.

Silicon.	Tin.

§ 61. According to the view we have taken, atomicity corresponds to complexity of atomic arrangement; and the elements of high atomicity consist of more vortex rings than those whose atomicity is low; thus high atomicity corresponds to complicated atomic arrangement, and we should expect to find the spectra of bodies of low atomicity much simpler than those of high. This seems to be the case, for we find that the spectra of Sodium, Potassium, Lithium, Hydrogen, Chlorine which are all monad elements, consist of comparatively few lines.

CAMBRIDGE: PRINTED BY C. J. CLAY, M.A. AND SON, AT THE UNIVERSITY PRESS.

3

Cavendish Professorship: First Years

Appointment as Cavendish Professor

In the autumn of 1884 Lord Rayleigh announced that he was resigning the Cavendish Professorship of Experimental Physics. Strenuous efforts were made to attract Lord Kelvin, the best known physicist of his generation, to the post. However, Kelvin was reluctant to leave Glasgow where he was well established.

In the event, five men applied for the position: Arthur Schuster, Thomson's former teacher and Professor of Applied Mathematics at Owens College; Osborne Reynolds, Professor of Engineering and also a former teacher of Thomson's; Richard T. Glazebrook, a demonstrator and lecturer at the Cavendish, with a great deal of experience in practical physics; Joseph Larmor, Thomson's old friend and rival who had beaten him to first place in the Mathematics Tripos and was now Professor of Natural Philosophy at Queen's College, Galway; and J.J. Thomson himself. Much to his surprise, Thomson was elected. 'I felt', he wrote, 'like a fisherman who with light tackle had casually cast a line in an unlikely spot and hooked a fish much too heavy for him to land.' [1]

The Cavendish community was also surprised by Thomson's election, Glazebrook having apparently been their favoured candidate. Thomson was only 28 and was not noted for his success as an experimental physicist. Yet it seems that, in some ways, this may have been in his favour. The practical research programme into electrical standards pursued by Rayleigh at the Cavendish was not universally popular in conservative college quarters, smacking too much of industrialism. Glazebrook's adherence to this programme probably weighed against him, while Thomson's theoretical research into the fundamental relations of ether and matter, his high ranking in the Mathematics Tripos, his Adams Prize, and his emerging theoretical reputation, counted strongly for him with the mathematical elite.

Of the other candidates, Schuster and Larmor were only marginally older, while Osborne Reynolds was more of an engineer than a physicist. None were currently established in Cambridge. Thomson had been a fellow of Trinity for four years and the tact and diplomacy which were to ensure the smooth running of the Cavendish for 35 years, and later to stand him in good stead as Master of Trinity College, were already apparent. Despite his middle-class origins, he identified himself completely with Cambridge ideals and patterns of behaviour. He was unlikely to upset the delicate balance between University and Colleges which was being threatened by the growth of University-funded science laboratories.

Thomson's election as Cavendish Professor was a turning point in his life in many ways. Hitherto he had been one promising physicist among many. Now, almost overnight, he became a leader of British science. His life was to be punctuated by requests to write reports (the *British Association Report on Electrical Theories* in 1885 being the first of these), to sit on committees, to edit journals, and to act as a spokesman for British science. To the man in the street, and the governing elite, these professional and administrative positions were probably more esteemed than his purely scientific work. Eventually they were to be rewarded by a knighthood in 1908, the Order of Merit in 1912, and his election as Master of Trinity College, a Crown appointment, in 1917.

As 28-year-old Cavendish Professor, Thomson was one of the most eligible young men in Cambridge. His salary of £500 per annum far outweighed that of College Fellows, who generally lived in College and had only recently been permitted to marry. In this small social circle, it was not long before Thomson met Rose Paget, daughter of Sir George Paget, Regius Professor of Physic (medicine). Rose was born in 1860, the elder of twin girls. She had little formal education, but was widely read, an intelligent listener and, according to her son George, had a passionate adoration of science, especially physics. She had acquired a lot of mathematics, attended some classes at Newnham College, and the elementary and advanced lectures and demonstrations at the Cavendish. In 1888 she was admitted as a research worker at the Cavendish, working on vibrations in soap films. A year later this research ceased abruptly when she and Thomson became engaged. They were married on 2 January 1890. As one of the few young couples in Cambridge, their home rapidly became the focus of a wide circle of friends, putting Thomson at the heart of Cambridge society. His father-in-law, George Paget, was a man of tremendous influence in the University, respected for his integrity and tact. From him, perhaps, Thomson learnt his way around the University administration, helping him steer the Cavendish unobtrusively through the difficult University versus College negotiations.

Gaseous Discharge

Even more important to Thomson, as a physicist, than marriage and honours, were the combined effects of job security and the requirement to be an experimental physicist. Hitherto he had identified his scientific work with the goals of the leaders of the Cambridge School of mathematical physics, men such as Kelvin, Rayleigh and Stokes. He had researched popular areas such as the vortex theory, Lagrangian mechanics and Maxwell's electrodynamics with great success, but little independence of judgement. He had started experimental research within Rayleigh's electrical standards regime, but this type of experiment, aimed at precision, did not suit him and he achieved no notable results. The theoretical papers he published between 1880 and 1884 were based on other people's experimental work, and it seems probable that he would have abandoned experimental physics entirely had he not been elected Cavendish Professor of Experimental Physics.

Now, however, he had an almost unassailable scientific position and a job for life. He had a duty to experiment, but was free to choose his own course. He was free even to choose to investigate a topic which Arthur Schuster recalled was deemed fit only for 'cranks and visionaries'. [2] This topic was electric discharge through gases. In it Thomson found an experimental programme that suited him down to the ground. He devoted the rest of his life to it and achieved outstanding success.

Gaseous discharge is the phenomenon seen when a glass vessel (a discharge tube) is filled with gas at low pressure and an electric potential is applied across it. The effects usually take the form of a glow in the tube, or fluorescence of the glass. They have been known since the seventeenth century, and are most familiar to us today as fluorescent light bulbs and television tubes. Gaseous discharge is, to quote Thomson, 'preeminent for the beauty and variety of the experiments and for the importance of its results on electrical theories'. [3] The effects depend considerably on the pressure of gas in the tube, the shape of the tube and electrodes, and the electric potential applied. The back cover of this book illustrates some of these effects.

The beauty and variety referred to exerted a strong hold on Thomson's imagination, and proved a constant source of inspiration to him. He was manually clumsy, which partly accounted for his previous failure as an experimentalist, but could now afford to pay an assistant to perform his experiments for him. However, he always sat by and watched, feeling he could 'see' what was going on in the gas. Had he investigated other areas of electromagnetism, such as electrolysis, he would have had to rely on meter readings for his results, whereas, '. . . in gases . . . the investigation of the electrical effects is facilitated by the visibility of the discharge, affording us ocular, and not merely circumstantial, evidence of what is taking place.' [4] Over the years he developed an extraordinary ability to see what was happening in the discharge tube and interpret it theoretically, as well as to diagnose what was wrong when the experiment did not work. E. V. Appleton, one of his later research students, recalled that, 'J.J., though quite innocent of manipulative skill . . . was . . . unique in his ability to conceive some new experimental method or some way of overcoming practical difficulties.' [5]

Thomson's enthusiasm for discharge experiments and the light they were throwing on his great theoretical questions saw him through the many frustrations that ensued. And frustrating these experiments often were to someone with his fertility of invention and eagerness to get on. He confided to Robin Strutt (later the Fourth Lord Rayleigh, and Thomson's first biographer) that he found delays and obstacles very trying when he was on fire with an idea that he wished to explore.

Thomson had strong theoretical reasons for choosing to work on discharge. He had first realised its possibilities during his work on vortex atoms for the Adams Prize in 1882, and the theory of discharge he consequently published in 1883 predated his election as Cavendish Professor. This was an entirely theoretical paper, but in broad outline the theory he developed guided his experiments for the next seven years. The paper is reproduced on p. 60.

According to his vortex theory of discharge, the chemical bonds holding atoms together in molecules were mediated somehow by the ether. Any cause, such as the electric field in discharge, which disturbed the ether might disrupt the bond and dissociate the molecule into its constituent atoms. Such dissociation of molecules into atoms should be apparent chemically, and the search for evidence of chemical dissociation with the passage of electricity through a gas became a recurring theme of Thomson's work. Indeed, the idea of dissociation guided his work on discharge for the rest of his life.

How, though, did dissociation explain the passage of electricity through the gas? According to Maxwell's theory, conduction of electricity was the sign of a relaxation of a strained state in the ether, which thus dissipated energy from the ether. Until 1890 or so, Thomson's main concern in his discharge theory and experiments was to find a mechanism for such energy dissipation.

Thomson's 1883 discharge theory was based on dissociation and association of vortex atoms and an analogy to electrolysis. Under certain conditions, if two vortex rings approach one another they remain together, circling about each other, for a length of time dependent on their energy and any external forces. This fact was shown by his mathematical analysis of vortices for the Adams Prize, and is shown visually in the plate opposite. Thomson took two such rings as his model for a molecule. He incorporated the hypothesis, due to Clausius and Williamson, that during electrolysis there is a constant interchange of atoms going on between molecules. He suggested that vortex rings were constantly pairing and re-pairing in the gas and introduced the ratio of paired time to free time. Following Maxwell, he treated the electric field as a velocity distribution in the ether, the vortices of which form the atoms. The velocity distribution disturbed the pairs and increased the free time. Since the energy of the two separate vortex atoms was higher than that of the vortex molecule, energy was absorbed from the ether in dissociating the molecules. Beyond a definite cut-off ratio of free time to paired time, Thomson regarded

Vortex molecules illustrated on Maxwell's Zoetrope. Fascination with the idea of vortex atoms was widespread in the 1870s and 80s. Maxwell had invented an improved form of zoetrope using lenses rather than slits to look through. He painted this special strip to go in it. The strip shows a molecule of three vortex rings constantly circling around each other. As the zoetrope is rotated the viewer can see (from left to right) a small fast moving vortex ring moving through the centre of two larger, slower ones. The fast moving ring slows and expands, as the first large ring contracts and moves, in turn through the other two. This is similar to the model that Thomson used for a molecule in 1883

J.J. Thomson and Rose Thomson, taken soon after their marriage

the gas as being dissociated; the atoms could be treated as fully independent and no energy was absorbed in increasing the free time further. The atoms might then recombine, emitting the surplus energy in the form of heat. This gave Thomson the mechanism he required for dissipating energy from the ether, and an explanation for conduction of electricity through gases in which dissociation of molecules (and hence chemical decomposition) was an integral part of discharge.

Thus discharge seemed to involve the three factors at the heart of Thomson's search for a unified theory: matter, electricity and the ether.

Thomson began experimenting on discharge in 1885, soon after his election as Cavendish Professor. At this time experimental work on discharge was very poorly regarded by the scientific establishment. No serious academic would touch it, and it was left in the hands of those for whom science was a sideline, such as Warren de la Rue and Hugo Muller, William Spottiswoode, and Henry Moulton, or those outside conventional academic circles whose livelihood relied on more spectacular science, such as William Crookes. This attitude was due to the lack of a coherent theoretical basis for investigation and to the very variety of effects. As Arthur Schuster, the only other academic interested in discharge, judged in 1884, 'the test of any theory must be found in the numerical results it is capable of giving. Hitherto, however, [in discharge] the qualitative phenomena have not, in my opinion, been sufficiently separated from the great number of disturbing effects to allow us to give a decisive value to quantitative measurements.' [6] Indeed, quantification in discharge research was virtually unheard of and results were generally reported as diagrams or photographs of the visual effects.

Thus, discharge did not lend itself to the approach of the predominant nineteenth-century Cambridge school of mathematical physics, dominated by such as Lord Kelvin, James Clerk Maxwell, Peter Guthrie Tait, George Gabriel Stokes, and Lord Rayleigh. In their *Treatise on Natural Philosophy*, the 'Bible' of this tradition, William Thomson (Lord Kelvin) and Peter Tait set out their 'rules for the conduct of experiments': 'When a particular agent or cause is to be studied, experiments should be arranged in such a way as to lead if possible to results depending on it alone; or, if this cannot be done, they should be arranged so as to show differences produced by varying it.' [7] Gaseous discharge was not amenable to such methods and was neglected accordingly.

Thomson, however, was not put off by such problems and the security of an established position enabled him to take on an unpopular research topic without fear of the consequences for his career.

In his work we can see an adaptation of the methods of the Cambridge school to the prevailing state of discharge research. He analysed his theories in great mathematical detail, using the approach of analytical dynamics. Yet, as we have seen, these methods, while they could predict trends and relationships, could not predict the size of effect to be looked for. Moreover, his theories were formulated in terms of vortices and strains in the ether, rather than observable experimental parameters. The only way Thomson could compare them with experiment was by analogy. He did this very clearly in one of his first experimental papers on discharge, in 1886, 'Some Experiments on the Electrical Discharge in a Uniform Electric Field, with Some Theoretical Considerations about the Passage of Electricity Through Gases', the first part of which is reproduced on p. 68.

This paper is a very interesting one for showing clearly the tensions between the Cambridge approach and the limitations of gaseous discharge. Thomson's concern to standardise his experimental conditions as much as possible stems from his Cambridge education. But his results, described verbally and by diagrams, show up clearly the prevailing ignorance of what the significant measurements were. The latter half of the paper chronicles a large number of known properties of discharge and shows how his theory could explain them, at least qualitatively and by analogy, but gives no indication of which of these properties were important and which were not.

This paper forms one of a series between 1885 and 1889, the main focus of which was the electric strength of gases. By this, Thomson meant the electric potential the gas could withstand before a discharge passed. According to his theory the electric strength was a measure of the cut-off ratio at which molecules dissociated permanently and discharge occurred. In it he might find a quantitative parameter for the theory of molecular dissociation and chemical reaction. He wanted the applied field as uniform as possible, so that he could be sure that his measured potential represented the true potential right across the discharge tube.

Thomson continued with this approach until 1889 but was unable to find any definite value of electric strength which could be attributed to a gas. The electric strength appeared to depend as much on the nature and condition of the electrodes as on the gas, and he concluded that it could not be a fundamental property (a conclusion that Schuster also reached in 1890). This conclusion, along with the influence of Hertz's recent experiments on electromagnetic waves, radically changed the direction of Thomson's theory and experiments in 1890, as we shall see in the next chapter.

The Changing Nature of the Cavendish

This very brief account of Thomson's work between 1884 and 1890 has glossed over the very great experimental difficulties he encountered in this work, and the fact that in undertaking it he began a process which was to change completely the nature of the Cavendish Laboratory.

Rayleigh, Thomson's predecessor, had instituted a programme of measurement of electrical standards in the Laboratory, a programme which exemplified the norms of the Cambridge mathematical physicists. For the preceding 20 years, British physicists had devoted much energy to trying to produce a reliable, reproducible and absolutely defined standard of electrical resistance. The significance of this was technological as well as theoretical. Resistance units were vitally important for submarine cable telegraphy, one of the foremost technologies of the period and one upon which control of the British Empire was coming to rely. Moreover, a precisely defined value of electrical resistance was also required to test Maxwell's increasingly powerful electromagnetic theory. In 1861 the British Association had formed a committee to produce such a standard. Maxwell was a member of the committee and in 1871, when he became the first Cavendish Professor, he arranged for the British Association equipment to be installed in the Cavendish. Rayleigh, succeeding Maxwell in 1879, stressed the electrical standards work and pursued it with outstanding success; it became the research programme by which the meaning of the Laboratory was defined. (Incidentally, Rayleigh himself achieved a high reputation for the reliability and accuracy of his experiments, which was later to pay off in the rapid recognition of his discovery of argon and his award of the Nobel Prize in 1904.) While not all the work of the Laboratory under Rayleigh was devoted to electrical standards, most was in a similar tradition; precise measurement of physical constants.

The successful physicist of the Rayleigh era had to be an accomplished mathematician, and virtually all research workers had taken the Cambridge Mathematical Tripos. Yet he needed a grasp of the physical significance of his mathematical manipulations. He required a high appreciation of experimental design, being aware of the capabilities and limitations of his apparatus, and must be good with his hands in order to make delicate adjustments. Above all, he must be patient and tenacious, for the experiments were fiddly and difficult to perform, yet the results were very limited. Imagination was not much called for. The skills of experimenting outside a well-established theoretical context, of developing theory and experiment hand in hand as Thomson had to in his discharge research, were unknown.

Thomson ushered in an era of tremendous excitement about physics. Experimental 'facts' were still all-important, yet the requirement for precision was gone. An order of magnitude was good enough for comparison with theory. The phenomena were not always well understood and the problems often ill-defined, but the results were nearly always highly significant. Imagination and a 'feel' for what was going on were stressed, the logic of mathematics providing subsequent justification. Researchers still had to be highly skilled with their hands, for most had to blow their own glass apparatus. Patience was still called for, now for tracing electrical connections and leaks in the vacuum systems.

These techniques and approaches of Thomson's took a long time to percolate through the Laboratory as a whole. Being young and inexperienced when he took over, Thomson was unwilling, or unable, to exercise the sort of benevolent despotism over the Cavendish that the aristocratic, landowning Rayleigh had done. He left the teaching in the hands of Rayleigh's experienced lieutenants, Richard Glazebrook and William Napier Shaw. Research workers were encouraged to choose their own topics and he made little attempt to draft them into his discharge programme. Only after 1895, when Cambridge threw open its doors to research students from other universities, did Thomson acquire a strong following within the Cavendish. Most of these men and women chose to come specifically to work with Thomson in his new and exciting field.

More practically, the techniques of discharge research were also unknown in the Laboratory, and Thomson had to acquaint himself, and the technicians, with them. He undoubtedly received a lot of help at the outset from his friend Richard Threlfall with whom he performed his first few discharge experiments. Threlfall was a highly skilled experimental physicist, who became Professor of Experimental Physics at the University of Sydney in 1886, rather to Thomson's dismay:

> 13 April 1886
> My dear Threlfall,
> . . . I cannot tell you how sorry I am you are going, every day I spend in the Laboratory when you are not there makes me feel that on that Monday when we elected you I did the worst days work for myself I ever did in my life, my only consolation is that I have established a permanent claim to the gratitude of New South Wales . . . [8]

Thereafter Threlfall and Thomson maintained a regular correspondence and these letters give us an excellent picture of Thomson's practical difficulties. Problems with pressure measurement, vacuum pumps and electrical sources are constant themes. Most outstanding though, were his problems with glassware and finding competent glass-blowers. Robin Strutt recalled

> Once when he came to me on his daily round and I unfolded a tale of woe, he said: 'I have been struggling with broken glass myself for a week past. I believe all the glass in the place is bewitched.' . . . In those days it was difficult to get glass of uniform composition. Comparatively few laboratories did any difficult glass-blowing and the suppliers of tubing did not turn over their stocks at all rapidly, nor did they always replenish them from the same source. The result was that much of the glass was deteriorated by age and one piece would not fuse satisfactorily to another of different composition.' [9]

Thomson was fortunate at the outset to appoint D.S. Sinclair as Laboratory instrument-maker and technician. Sinclair had a wide range of skills, was a good chemical technician, and an excellent glassblower. He was crucial to Thomson's early work:

> June 13 1886
> Dear Threlfall,
> . . . One experiment which I have got through with has been to try the effects of big sparks through a long tube in which there were nothing but fused glass joints. We made the nitrogen by heating ammonium bichromate in a long tube, as N comes off too quickly to be purified from the bichromate by itself we mixed it with very pure silica – this tube was fused on to

some purifying bulbs containing HCl, KHO and H_2SO_4 after these came a tube of P_2O_5 this was sealed on to the discharge tube and the discharge tube then sealed on to the pump. Sinclair had a very neat dodge for avoiding the difficulty caused by the liquids in the bulbs of making the last joint. He fused in two barometer tubes and then blew through these. They afterwards dipped into Hg and served as additional gauges. The whole thing when done was a great achievement in glass blowing and measured nearly 12 feet in length . . . . [10]

However, Sinclair left the Cavendish in December 1886, to set up his own business. His replacement, A.T. Bartlett, was unable to blow glass, and Thomson's discharge work came to a complete standstill for several months.

February 3 1887
Dear Threlfall,
. . . It is a long time since I wrote to you but I have been having a lull in my experiments as I have been stopped for want of a glass blower since Sinclair left. I have been having the new man taught and I think he will be able to do it all right . . . [11]

March 20 1887
Dear Threlfall,
. . . I feel Sinclair's loss very much as the new man is not worth a cent as a glass blower and I have to pay a boy in the Chemical lab. to blow my tubes . . . [12]

August 7 1887
Dear Threlfall,
. . . I am going to get a private assistant who can do glass blowing as we feel the want of a glass blower very much. My sparking experiments were stopped for some time whilst I was waiting for some tubes from London, which however have turned up at last . . . [13]

At length, in frustration, Thomson poached from the Chemistry laboratory an assistant, Ebeneezer Everett, who had taught himself the art of glass-blowing.

September 4 1887
Dear Threlfall,
. . . I have got 'Ebeneezer' from the Chemical Lab as my private assistant. I took him chiefly because he is a good glass blower and we have no one in the Laboratory who is worth a rush at it . . . [14]

Thomson paid Everett out of his own money as his own personal assistant. Everett rapidly became central to the discharge research programme and worked for Thomson until 1930 when he had to retire due to ill health. Everett was devoted to Thomson's interests, but relations were not always harmonious as Robin Strutt recalled:

> Sometimes when things were not going well a little tension was apparent in the relations between J J and Everett; Everett being convinced that J J's view of what should occur was wrong, and J J sure that the failure of the experiment was due to Everett not having carried out his instructions properly. On one occasion J J was attempting to deflect by means of a magnetic field positive rays produced in a glass globe some six inches in diameter, situated between the poles of a large Du Bois electromagnet. J J had his eye to a microscope and was observing the motion of the gold leaf of an electroscope across a scale. The conversation was overheard.
>
> J J: 'Put the magnet on.' Then followed a click as Everett closed a large switch.
> J J: 'Put the magnet on.'
> Everett: 'It is on.'
> J J (eye still to the microscope): 'No it isn't on. Put it on.'
> Everett: 'It is on.'
>
> A moment later J J called for a compass needle. Everett went out of the room and returned with a large needle, 10 inches long, which was used in elementary lectures in magnetism. J J took it, and approached the electromagnet. When about a foot away the needle was so strongly attracted by the electromagnet that it swung round and flew off its pivot, crashing into the bulb (which burst with a loud report) and coming to rest between the poles of the magnet. The spectators looked up to see what had happened. Everett was glowing with triumph, and J J looking at the wreck with an air of dejection. 'Hm,' he said. 'It was on.' [15]

In 1933, Everett died and Thomson wrote his obituary in *Nature*:

> Everett took a very active and important part in the researches carried on in the Laboratory, by students as well as by the professor. The great majority of these involved difficult glass blowing, which was nearly all done by Everett, as it was beyond the powers of most of the students. In addition to this, he made all the apparatus used in my experiments for the more than forty years in which he acted as my assistant. I owe more than I can express to his skill and the zeal which he threw into his work. He was a very skilful glass blower, a quick worker, very pertinacious; if the first method failed he would try another and generally succeeded in finding one which would work. He was also an excellent lecture assistant, and was a great help to me in my lectures at the Royal Institution . . . In the early days of X-rays, before hospitals or medical men had any appliances for taking X-ray photographs, Everett and W.H. Hayles, another assistant at the Cavendish Laboratory and an expert photographer, organised a scheme for taking photographs at the Laboratory. Many medical men availed themselves of this, although the revelations made by the photo-

The Cavendish Laboratory

Laboratory assistants 1900. Everett is on the front right. W.H. Hayles (back left) helped Thomson extensively with photographic work. F. Lincoln (back, second from right) became a longstanding laboratory superintendent

graphs as to the way in which bones had been set sometimes caused considerable embarrassment. [16]

References

Cambridge University Library holds an important collection of Thomson manuscripts, classmark ADD 7654, referred to here as CUL ADD 7654, followed by the particular manuscript number. Other archives, and a more complete bibliography, may be found in: Falconer, I., Theory and Experiment in J.J. Thomson's work on Gaseous Discharge, PhD Thesis, 1985, University of Bath.

[1] THOMSON, J. J., *Recollections and Reflections*, 1936, London: Bell. Reprinted 1975, New York: Arno, p. 98.
[2] SCHUSTER, A., *Progress of Physics During 33 Years*, 1911, Cambridge University Press, p. 52.
[3] CUL ADD 7654 BD2.1.
[4] THOMSON, J. J., Electric Discharge Through Gases. *Notices of Proceedings of the Royal Institution*, **14** (1894) 239.
[5] APPLETON, E. V., Obituary of Thomson. *Nature*, **146** (1940) 353.
[6] SCHUSTER, A., Experiments on the discharge of electricity through gases: sketch of a theory. *Proceedings of the Royal Society*, **37A** (1884) 318.
[7] THOMSON, W. AND TAIT, P. G., *Treatise on Natural Philosophy*, 2nd edn, 1912, London, pp. 442–443.
[8] CUL ADD 7654 T7.
[9] RAYLEIGH, FOURTH LORD, *The Life of Sir J.J. Thomson*, 1942, Cambridge University Press. Reprinted 1969, London: Dawsons, p. 25. Page references are to the later edition.
[10] CUL ADD 7654 T9.
[11] CUL ADD 7654 T15.
[12] CUL ADD 7654 T16.
[13] CUL ADD 7654 T19.
[14] CUL ADD 7654 T20.
[15] See reference [9], p. 152.
[16] THOMSON, J. J., Obituary of Ebeneezer Everett. *Nature*, **132** (1933) 774.

LXIV. *On a Theory of the Electric Discharge in Gases. By* J. J. THOMSON, *M.A., Fellow of Trinity College, Cambridge**.

THE aim of the following article is to give an account of a theory which seems to explain some of the more prominent phenomena of the electric discharge in gases, and which also indicates the presence in the electric field of stresses consisting of tension along the lines of force combined with pressures at right angles to them. Maxwell, as is well known, showed that stresses of this character would explain the mechanical actions between electrified bodies.

I shall take the vortex-atom theory of gases as the basis of the following remarks, as it possesses for this purpose advantages over the ordinary solid-particle theory; though much of the reasoning will hold whichever theory of gases be assumed. As the vortex-atom theory of gases is not very generally known, I shall begin by quoting the more important consequences of this theory which are required in this article. According to this theory, the atoms of gases consist of approximately circular vortex rings. When two vortex rings of equal strength, with (as we shall suppose for simplicity) their planes approximately parallel to each other and approximately perpendicular to the line joining their centres, are moving in the same direction, and the circumstances are such that the hinder ring overtakes the one in front, then if, when it overtakes it, the shortest distance between the circular axes of the rings be small compared with the radius of either ring, the rings will not separate, the shortest distance between their circular axes will remain approximately constant, and these circular axes will rotate round another circle midway between them, while this circle moves forward with a velocity of translation which is small compared with the linear velocity of the vortex rings round it. We can prove that in this case the product of the momentum and velocity of the rings is greater than the sum of the products of the same quantities for the rings when they were separated by a distance great compared with the radius of either. We may suppose that the union or pairing in this way of two vortex rings of different kinds is what takes place when two elements of which these vortex rings are atoms combine chemically; while, if the vortex rings are of the same kind, this process is what occurs when the atoms combine to form molecules. If two vortex rings paired in the way we have described are subjected to any disturbing influence, such as the action due to other vortex rings in their neighbourhood, their radii will be changed

* Communicated by the Author.

by different amounts; thus their velocities of translation will become different, and they will separate. We are thus led to take the view of chemical combination put forward by Clausius and Williamson, according to which the molecules of a compound gas are supposed not to always consist of the same atoms of the elementary gases, but that these atoms are continually changing partners. In order, however, that the compound gas should be something more than a mechanical mixture of the elementary gases of which it is composed, it is evidently necessary that the mean time during which an atom is paired with another of a different kind, which we shall call the paired time, should be large compared with the time during which it is alone and free from other atoms, which time we shall call the free time. If we suppose that the gas is subjected to any disturbance, then this will have the effect of breaking up the molecules of the compound gas sooner than would otherwise be the case. It will thus diminish the ratio of the paired to the free time; and if the disturbance be great enough, the value of this ratio will be so much reduced that the substance will no longer exhibit the properties of a chemical compound, but those of its constituent elements: we should thus have the phenomenon of dissociation or decomposition.

The pressure of a gas in any direction is, according to the vortex-atom theory of gases, proportional to the mean value of the product of the momentum and velocity in that direction, just as in the ordinary solid-particle theory.

Let us now suppose that we have a quantity of gas in an electric field. We shall suppose, as the most general assumption that we can make, that the electric field consists of a distribution of velocity in the medium whose vortex-motion constitutes the atoms of the gas; the disturbance due to this distribution of velocity will cause the molecules of the gas to break up sooner than they otherwise would do. Thus the ratio of the paired time to the free time will be diminished. Now, when the atoms are paired, the product of the momentum and velocity for the compound molecule is greater than the sum of the products of the same quantities for the constituent atoms when free, but the pressure in any direction is proportional to the mean value of the product of the momentum and velocity in that direction. Thus each atom will contribute more to the pressure when it is paired than when it is free; and thus, if the ratio of the paired to the free time be diminished, the pressure will be diminished. Now, according to any conception which can be formed of the distribution in the medium of the velocity due to the electric field, the variation in the velocity will be greater along the lines of force than at right angles to them; and thus those molecules which are moving along the lines of force will be split up into atoms sooner than those moving at right angles to them. Thus the ratio of the paired to the free time will be less for those molecules which are moving along the lines of force than for those moving at right angles to them; and therefore the pressure will be less along the lines of force than at right angles to them. Maxwell, in his 'Treatise on Electricity and Magnetism,' has shown that a distribution of stress of this character will account for the mechanical actions between electrified bodies.

To show that it is conceivable that this cause should produce effects sufficiently large to account for electrostatic attractions and repulsions, it may be useful to point out that the electric tension along the lines of force is very small when compared with the atmospheric pressure; for air at the atmospheric pressure the maximum electric tension is only about $\frac{1}{2000}$ of the atmospheric pressure. The theory that electrostatic attractions and repulsions are due to stresses in the gaseous dielectric admits of an experimental test; for it is evident that, according to this theory, the tension along the lines of force cannot exceed the pressure of the gas. Thus, if we have a gas sufficiently rare to support an electric field so intense that the pressure of the gas does not greatly exceed the electric tension along the lines of force when calculated by the ordinary expression, viz. $KH^2/8\pi$ (where K is the specific inductive capacity of the gas, and H the electromotive force), then, if the theory of stress in the gas be correct, the tension along the lines of force will soon reach a maximum value, and will not increase with an increase in the electromotive force. Thus the attraction between the two electrodes in this case would reach a maximum, and would not afterwards increase with an increase in the difference of potential between them. I may point out that, for this to happen, the density of the gas would have to be much less than the density for which the electric strength is a minimum, which the researches of Dr. De La Rue and others have shown to be at a pressure about ·6 millimetre. For down to this pressure the electric force necessary to produce discharge is, speaking very roughly, proportional to the pressure, but the electric tension is proportional to the square of the electromotive force. Thus, down to the pressure of minimum strength, the ratio of the greatest electric tension to the pressure of the gas diminishes with the pressure; and it would be no use seeking for any effect such as is described above, except at pressures very much less than this. Taking the formula given by Dr. Macfarlane in the Philosophical Magazine for December 1880, for calculating the electromotive force necessary to produce discharge from the pressure when this is less than that giving the minimum electrical strength, viz. $V = \cdot 67/\sqrt[3]{p}$ (where V is the difference of potential per centimetre, and p the pressure in millimetres of mercury), I find that at a pressure of about ·0001 of a millimetre of mercury the electric tension just before discharge would equal the pressure of the gas; so that it is at pressures comparable with this that the experiment ought to be tried.

Let us now pass on to the case where the intensity of the electric field is so great that the dielectric can no longer insulate, and the electricity is discharged.

It will be instructive to consider for a moment what happens when a compound gas is raised to such a temperature that it is dissociated, or an elementary one until its molecules are split up into atoms. If the gas is at a low temperature, say 0° C., when heat is first applied, so far as we can tell the whole of the heat is employed in raising the temperature and increasing the radiation, and no heat is rendered latent; in other words, the alteration in the molecular structure of the gas absorbs no work. This state of things continues until we approach the temperature at which the gas begins to be dissociated; then a large fraction of the heat supplied to the gas is used up in altering the molecular structure, and only a part of it is spent in raising the temperature and increasing the radiation.

If we look on this from the point of view of chemical combination which we took before, we may regard it as showing that, if any energy be supplied to the gas when the ratio of the paired to the free time is so large that the gas exhibits none of the phenomena of dissociation, the consequent diminution in the ratio of the paired to the free time does not absorb any of the energy; but if the ratio of the paired to the free time be so small that the gas exhibits some of the phenomena of dissociation, then a diminution in the ratio of the paired to the free time will absorb a considerable amount of energy. The same statements will apply to an elementary gas, except that in this case the change of structure consists in splitting the molecules up into atoms of the same kind, while in the compound gas they were split up into atoms of different kinds.

Let us now apply these considerations to the case of the electric discharge. The disturbance to which the gas in an electric field is subjected makes the molecules break up sooner into atoms than they otherwise would do, and thus diminishes the ratio of the paired to the free times of the atoms of the gas; as the intensity of the electric field increases, the disturbance in some places may become so violent that in these regions the ratio of the paired to the free times approaches the value it has when the gas is about to be dissociated. At this point any diminution of this ratio consequent upon an increase in the intensity of the field will absorb a large amount of energy; this energy must come from the electric field; and we should thus get the phenomenon of the electric discharge. The disturbance to which the gas is subjected might very well account for the luminosity of the discharge; whilst the heat produced by the recombination of the dissociated gas, which would occur as soon as the disturbance due to the electric field was withdrawn, would account for the heat produced by the discharge. Only a very small amount of gas would have to be decomposed in order to absorb the electrical energy of the field. Taking the values of the electric force necessary to produce discharge given by Dr. A. Macfarlane (Phil. Mag. Dec. 1880), we find that if the dielectric be hydrochloric-acid gas and the gaseous layer be a centimetre thick or more, the electric energy per cubic centimetre will be less than 1000 in C.G.S. units; while the amount of energy necessary to decompose 1 cub. centim. of hydrochloric-acid gas is more than 4×10^9 in the same units. Thus only about one four-millionth part of the gas would have to be decomposed in order to exhaust the energy of the electric field. This quantity, though so small, is yet probably much larger than would in reality be required, as the work which absorbs the energy of the electric field is the splitting-up of the molecules of the hydrochloric-acid gas into hydrogen and chlorine atoms. These atoms will combine with each other to form molecules of hydrogen and chlorine respectively, and will give out heat in so doing, which will heat the gas, but will not restore the electric energy. Now the heat of combination of hydrogen and chlorine when they form hydrochloric acid is not the same as the heat required to split up hydrochloric acid into atoms of hydrogen and chlorine, but is equal to the latter quantity minus the heat given out when the atoms of hydrogen and chlorine combine to form molecules of hydrogen and chlorine respectively. The determinations by Prof. E. Wiedemann of the heat given out when hydrogen atoms combine to form molecules, and by Prof. Thomsen of the same quantity for carbon atoms, seem to show that these quantities are greater than the heat given out in ordinary chemical reactions, and thus that these latter quantities are the differences of quantities each much greater than themselves. Thus to decompose hydrochloric acid into hydrogen and chlorine would require much more energy than the mechanical equivalent of the heat of combination of hydrogen and chlorine.

This view of the electric discharge indicates a relation between the electric strength of a gas and its chemical properties; for in order to make the spark pass through an elementary gas we have to decompose the molecules into atoms: thus the stronger the connexion between the atoms in the molecule, the greater the electric strength. Thus, for example, we should expect that the atoms of nitrogen are much more firmly connected together in the molecule than the atoms of hydrogen, as the electric strength of nitrogen is much greater than that of hydrogen. Unfortunately we seem to know very little about the strength of connexion between the atoms in the molecule; it would, however, be interesting to try whether a gas whose molecules, like those of iodine vapour, are easily dissociated into atoms would be electrically weak. In many cases, of course, the decomposition of the gas on the passage of the spark is very evident; a common way of decomposing a gas being to pass sparks through it. We might, however, have chemical decomposition without being able to detect the products of it; for these might recombine as soon as the disturbance produced by the electric field was removed. In the case of an elementary gas, the splitting-up of the molecules into atoms will effect the same purpose as the decomposition of the compound gas—*i. e.* the exhaustion of the electric field. Thus, according to the view we are now discussing, chemical decomposition is not to be considered merely as an accidental attendant on the electrical discharge, but as an essential feature of the discharge, without which it could not occur.

Let us now consider what effect rarefying the gas would have upon its electrical strength. In a rare gas the mean distance between the molecules is greater than in a dense one; and if the temperature be the same in both cases, and consequently the mean velocity of the molecules the same, the ratio of the free to the paired time will be greater for the rare than for the dense gas; for the free atoms will, on an average, be longer in meeting with fresh partners. Thus the rare gas will be nearer the state in which it begins to suffer dissociation than the dense gas, and thus it will not require to be disturbed so violently as the dense gas in order to increase the ratio of the free to the paired time to its dissociation value; and thus the intensity of the field necessary to produce discharge would be less for the rare gas than for the denser one: in other words, the electric strength would diminish with the density; and this we know is the case. It is now generally admitted that rare gases are more easily dissociated than denser ones. In fact Sir C. W. Siemens takes this as the basis of his theory of the Conservation of Solar Radiation, as he supposes that the rays of the sun are able to dissociate the compound gases, chiefly hydrocarbons, which in a very rare state he supposes distributed throughout the universe; while, when these gases exist at pressures comparable with that of the atmosphere, they are able to transmit the sun's rays without suffering dissociation. These considerations would seem at first sight to indicate that the electric strength of gases would continually decrease with the density; whereas we know that it only does so to a certain point, and that afterwards the electric strength increases as the density decreases.

We have in the above reasoning, however, supposed that whenever we got chemical decomposition at all we had always sufficient energy absorbed to exhaust the electric field. In consequence of the great absorption of energy in chemical decomposition, this is legitimate, unless the gas be very rare; but for a very rare gas it will be necessary to decompose a larger proportion of the molecules of the gas, and it will require a more intense electric field to do this. If the gas were very rare, it might be that the energy required to decompose all the gas was not sufficient to exhaust the energy of the electric field. In this case all the electricity could not be discharged at once; while in an absolute vacuum there would be no chemical decomposition to lessen the energy of the electric field, and there would be no electrical discharge at all. Thus there are two causes at work which produce opposite effects on the electric strength as we rarefy a gas. The first is that the gas is more easily dissociated as we rarefy it; this diminishes the electric strength of the gas. The second is that, as there are fewer molecules, a larger proportion of them must be decomposed in order to exhaust the same amount of energy, and it will require a more intense electric field to separate the larger proportion; this will tend to increase the electric strength of a gas as we rarefy it. The second of these considerations is not important at pressures comparable with that of the atmosphere, as in this case the percentage of the gas which has to be dissociated in order to exhaust the energy of the electric field is extremely small; so that, starting from the atmospheric pressure, we should expect the gas to get electrically weaker as it gets rarer. With very rare gases, on the other hand, the second consideration, as the extreme case of a perfect vacuum shows, is the more important; and thus with very rare gases the electric strength should increase as the gas gets rarer. Both of these results agree with the results of experience.

It may be worth while to point out that, according to the view taken in this paper, a perfect vacuum possesses infinite electric strength; and thus it is in opposition to the theories put forward by Prof. Edlund and Dr. Goldstein, in both of which a vacuum is regarded as a perfect conductor.

In a future paper I hope to explain other phenomena of the electric discharge by means of this theory, and also to apply it to the case of conduction through metals.

[Extracted from the *Proceedings of the Cambridge Philosophical Society*, Vol. v. Pt. vi.]

(1) *Some experiments on the electric discharge in a uniform electric field, with some theoretical considerations about the passage of electricity through gases.* By Professor J. J. THOMSON.

As the experiments which have hitherto been made on the discharge of electricity through gases have in general been arranged in such a way that it is difficult to calculate what was the state of the electric field before discharge took place, I have thought it might be interesting to make some experiments when the state of the field was accurately known. For this reason I made the discharge take place between two parallel plates separated by a distance which was but small in comparison with their diameters.

Fig. 1.

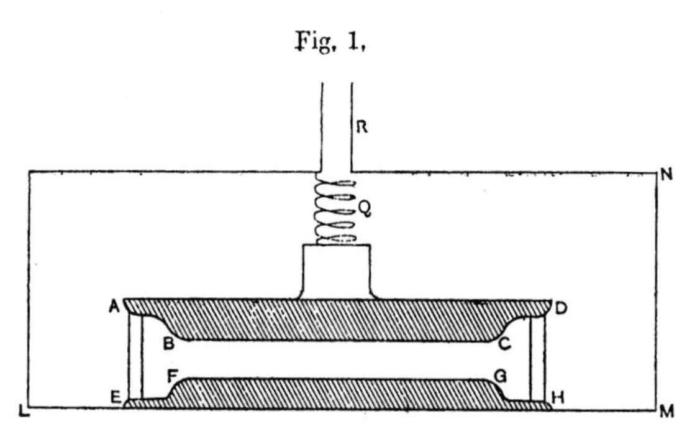

The arrangement used is represented in fig. 1. *ABCD, EFGH* are two cast-iron plates, the flat portions of which are about 6 centimetres in diameter, and 1½ centimetres apart. They are shaped as in the figure, special care being taken to make the curved parts of the plate smooth and free from places of large curvature; the object of this as well as the peculiar shape of the electrodes is to make the electric field much less intense in those places where it is not uniform than in those places where it is, so that the discharge will take place in the uniform field and not in those places where the field is not uniform and difficult to calculate. The surfaces of the plates were worked very true, and some small holes that were left in from the casting were filled up with putty and then coated with gold leaf. The surfaces were so true that though the electrodes were of considerable weight, yet if they were placed in contact they adhered sufficiently to cause the under one to be lifted when the upper one was raised. The plates were maintained at the same distance apart by means of three glass distance-pieces, two of which are shewn at *AE* and *DH*, carefully made of the same length and their ends accurately ground; these were connected together by pieces of glass rod: these distance-pieces were placed in the hollow part of the plates so as to be out of the way of the discharge; the plates were placed in a box, *LMNP*, the side of which was a cylindrical piece of glass and the ends of it brass discs, fastened to the glass with marine glue; into the upper one of these plates a piece of brass tubing, *R*, was soldered in order to permit of the exhaustion of the gas in the vessel; between the top of the box and the upper plate there was a spring, *Q*, which put the two into electrical connection. The spark was produced by means of an induction coil.

The following are the phenomena which occur as the air is gradually exhausted from the box. At the pressure of the atmosphere the spark passes between two points, being evidently determined by some accident which makes the force a little greater at one place than another; at this stage the discharge is very unsteady and skips about from one point of the plates to another: as the pressure diminishes the discharge gradually settles down and remains at one place, and begins to present peculiar features which are represented in the accompanying figures, in all of which the negative electrode is supposed to be at the top. Fig. 2 re-

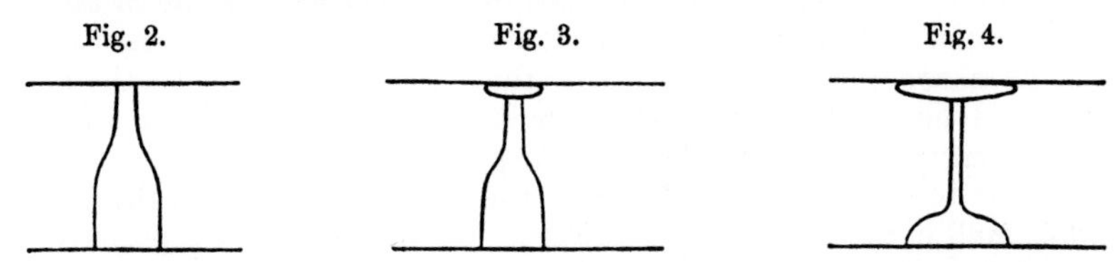
Fig. 2. Fig. 3. Fig. 4.

presents the discharge when the pressure is that due to about 90 mm. of mercury; it is shaped something like an Indian club with the handle at the negative electrode. As the pressure diminishes, the neck of the club lengthens, the lower part broadens out, and a disc appears at the negative electrode; the appearance at a pressure of about 40 mm. of mercury is represented in fig. 3; the discharge being bluish near the negative electrode, but reddish towards the positive. As the pressure falls the disc near the negative electrode broadens out, and the handle of the Indian club lengthens, until the appearance is that represented in fig. 4, which represents the discharge at about 18 mm. of mercury, the difference in colour between the discharge at the positive and negative electrodes being now very marked. As the pressure diminishes the disc at the negative electrode increases in size, until at about 4 mm. this disc appears to constitute the whole of the discharge; it is clearly separated from the negative electrode. I have not been able to detect with any certainty any discharge at the positive electrode, or any glow throughout the tube, and if they exist at this stage they are certainly exceedingly faint. There is a much greater contrast between the bright disc near the negative electrode and the rest of the discharge than between the glow and dark space in a vacuum tube of the ordinary kind. As the pressure diminishes still further, the disc gradually moves further away from the negative electrode, and a decided glow spreads through the vessel; the colour of the discharge keeps changing, and when the pressure sinks below a millimetre it is a pale Cambridge blue. Bright specks also appear over the negative electrode. If an air-break be put in the circuit a curious phenomenon is observed. A glow is distinctly visible between the top of the vessel containing the electrodes and the upper electrode, though these are in metallic connection, and if they acquired the same potential simultaneously, could have no electric field between them. This was only observed when the upper electrode was negative, not when it was positive. The lowest pressure reached with this apparatus was about $\frac{1}{5}$ of a millimetre. At this pressure the disc near the negative, though still observable, was not much brighter than the surrounding glow.

These experiments were repeated, using coal-gas instead of air: very similar results were obtained, except that at high pressures the discharge jumped about more than it did in air. Coal-gas was used because stratifications are usually produced in it with great facility. I never observed any tendency however in the discharge to become striated where the field was uniform, though some small discharges which started from the edges of the positive plates were beautifully striated; on one occasion too a spot of dust had got on the positive electrode in the middle of the uniform discharge, a secondary discharge started from this point, which was very plainly striated, though the main discharge shewed no trace of stratification.

Some experiments were tried with gases which are electrically very weak, such as the vapours of turpentine and alcohol; with these gases I never was able to limit the discharge to a disc near the negative electrode, as was the case when air or coal-gas was

28—2

used. With these vapours there was always at low pressures a glow stretching across the space between the electrodes, and though the disc near the negative electrode was distinctly brighter than the rest, I was never able as in air to get the discharge practically confined to the disc. The brightness of the glow was always comparable with that of the disc.

The stratifications of the discharge followed the same law as in coal-gas; in the parts of the discharge where the field was uniform no tendency to stratification could be detected, but any secondary discharge started by some accidental inequality was always distinctly striated.

With the arrangement described above, the pressure could not

Fig. 5.

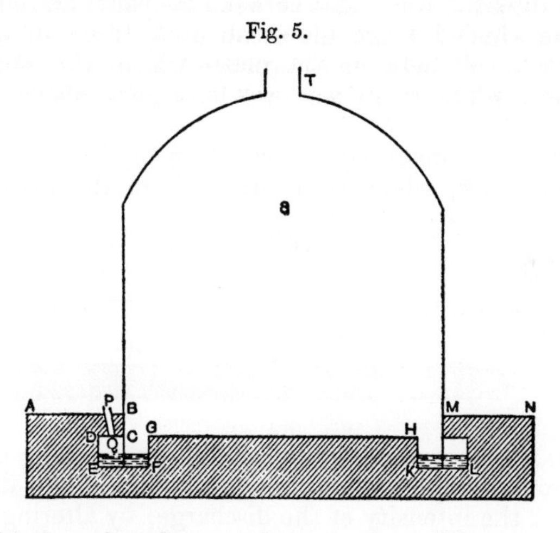

be reduced below that due to about $\frac{1}{5}$ of a millimetre of mercury, and it was with great difficulty that it was reduced as low as this, so in order to investigate the phenomena at higher exhaustions the parallel plates were put in the apparatus represented in section in fig. 5.

ABCDEF is a brass bed plate with a groove of the shape *CDEFG* cut in it; into this groove the glass vessel *S* which terminates in the tube *T* fits, and is fastened against the side *BC* of the groove by marine glue; mercury is poured into the groove, and the space outside the glass vessel between the mercury and the brass is kept exhausted by a Sprengel pump, which is connected with this space by means of the tube *PQ*. Thus since the pressure in this space can easily be kept by the Sprengel lower than the pressure due to 5 millimetres of mercury, if the depth of the mercury in the groove be greater than 5 millimetres, no air can pass into the vessel, however low the pressure inside may be. The glass tube T was fused on to a Töpler pump. With this arrangement the only joints between the inside of the vessel S and the outside air are mercury ones, and the vessel can readily be exhausted.

The discharge presents the following appearance as the vessel is gradually exhausted. After the pressure gets below the value reached with the first arrangement the glow between the plates gets more and more uniform, until no difference can be perceived in the intensity of the light between the plates; as the pressure diminishes a glow appears above the upper plate, of the kind to be described below, and the intensity of the glow between the plates diminishes; as the pressure falls still lower the glow above the plates increases in intensity while that between the plates diminishes, until at a pressure which I estimated at about $\frac{1}{50}$ of a millimetre there was no glow at all between the plates which were separated by a dark space, while there was a strong glow above the upper plate.

This glow is represented in section in fig. 6. *ABCD, EFGH* represent the glow, which is separated from the plate *LMN* by

Fig. 6.

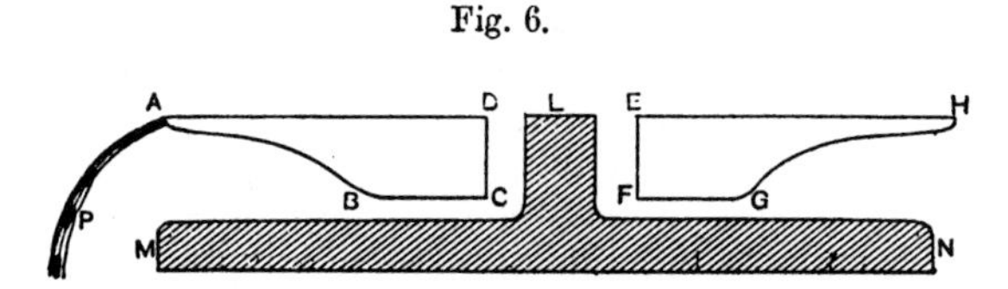

dark spaces, which are left blank in the figure. The distance of the glow from the plates, that is the width of the dark space, depends upon the intensity of the discharge; by altering the screw of the commutator of the coil the glow could be made to rise and fall in a very striking fashion. The stronger the discharge the smaller seemed the dark space.

From one edge of the glow a bright thread of striated glow, *AP*, extends, forming apparently the positive part of the discharge, the glow being the negative; this positive part looked like a continuation of the negative, there being no interval that I could see between the glow and the striated discharge. This striated discharge started from the place where the negative glow was farthest from the glass of the bell-jar. It was extremely sensitive to the action of a magnet, the point from which the discharge started being altered by the magnet. The direction in which the discharge moved was along the circumference of the glow, and the direction was determined by the component of the magnetic force along the radius from the centre of the glow to the point where the discharge took place; if the direction of this component was reversed the direction of displacement of the glow was reversed. When this component was in one direction the striated discharge was not only deflected but split up into several discharges, there being in this case often 7 or 8 striated discharges proceeding from the negative glow; when the direction of the magnetic force was reversed, the discharge was deflected in the opposite direction, and instead of being split up seemed to be more concentrated than before. This part of the effect seemed to be due to the action of the magnet on the glow; the place where the striated discharge starts is where the glow is furthest from the glass; if the magnetic force by its action on the glow reduces the inequalities in the distance of the edge of the glow from the glass the discharge may start from several places at once, while if it tends to increase the inequalities the glow will be more rigorously confined to one place.

Theoretical considerations about the electric discharges in gases.

In a paper published in the *Philosophical Magazine* for June, 1883, page 427, I gave a theory of the electric discharge in gases, in which the discharge was regarded as the splitting up of some of the molecules of the gas through which the discharge takes place; the energy of the electric field being spent in decomposing these molecules, and finally by the heat given out on the recombination of the dissociated atoms appearing as heat, except in the numerous cases where the gas is permanently decomposed by the spark, when part of the energy of the field remains as potential energy of dissociated gas.

In that paper I did not discuss the difference between the effects observed at the positive and negative electrodes. I think however that the theory is capable of explaining these differences. For we may imagine a molecule of such a kind that the atoms in it would tend to separate when the molecule was moving in one direction in an electric field, say that of the lines of force, but would be pushed nearer together when the molecule was moving in the opposite direction. A molecule of the following kind would possess this property.

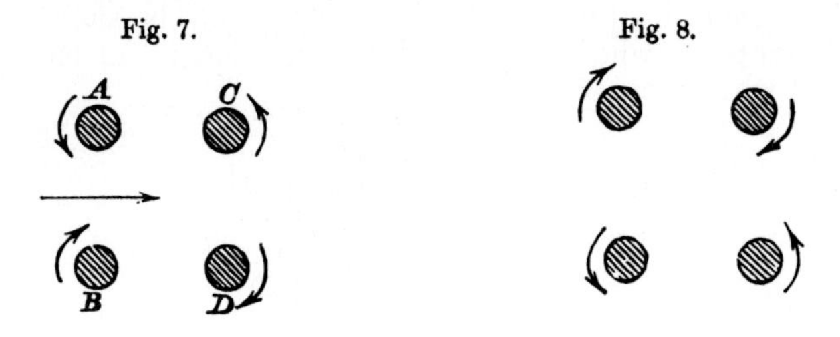

Suppose we have two vortex rings *AB* and *CD* of equal strength, whose planes are parallel and whose cores are nearly coincident, they will rotate round each other, the cores remaining at an approximately constant distance apart. Let us suppose that these rings are moving in a fluid which is in motion but in which the distribution of velocity is not uniform; then we know (see a Treatise on the 'Motion of Vortex Rings' by J. J. Thomson, p. 65) that the radii of the rings will alter, and that the alteration will not be affected by reversing the direction of motion of the rings.

Now let us suppose that the radius of *AB* in consequence of the distribution of velocity in the surrounding fluid increases faster than that of *CD*, then since the velocity of a ring diminishes as its radius increases the diminution in the velocity of *AB* will be greater than in that of *CD*, so that *CD* will now move faster than *AB*, the distance between the rings will therefore increase, and if the difference between the velocities is great enough they will ultimately separate. Next let us suppose that the rings are turned round so as to be moving in the opposite direction, as in fig. 8. Then, since the alteration in the radius of either ring is the same after the direction of motion has been reversed; under the same circumstances as before, the radius of the ring *AB*, which is now in front, will still increase faster than that of the ring *CD*, which is now in the rear; that is, the diminution in the velocity of the ring in front will be greater than that of the one in the rear, that is, the front ring will move more slowly than the one behind, so that the distance between the rings will diminish and the connection between the atoms in the molecule be made firmer, while in the other case the molecules tended to separate. The only difference between the cases, however, is the direction in which the molecules are moving, so that a molecule of this kind may tend to be decomposed when it is moving in one direction and not when it is moving in the opposite one.

It would, I have no doubt, be possible to give an illustration of this property by taking an ordinary mechanical system and supposing it to be acted on by a proper distribution of forces: the above illustration, however, is sufficient for my purpose, which is to shew that the properties of molecules may be such that they are decomposed when moving in one direction in an electric field but not when moving in the opposite.

Let us trace some of the consequences of supposing that the molecules are decomposed when moving in the direction of the lines of force and not when moving in the opposite direction. If we consider the electric field near the electrodes, this means that at the negative electrode those molecules which are moving towards it are the only ones which have any tendency to be decomposed, while at the positive electrode it is only those molecules which are moving away from it which are in this condition.

The consequence of this will be that the molecules will be more easily decomposed at the negative than at the positive electrode. For consider first of all the case of a non-uniform field, when the intensity of the field diminishes as we recede from the electrodes. At the negative electrode those molecules which are approaching the electrode are the ones which tend to get decomposed, and these are going from weak to strong parts of the field, so that the tendency to dissociate gets stronger and stronger, while it keeps getting a better leverage, as it were, for the atoms in the molecule get further and further apart as the molecule moves, and thus the difference in the alteration in their radii would increase even if the field were uniform, but when the field increases in intensity, as the molecule moves on, the effect is still more increased. On the other hand, those molecules at the positive electrode which are likely to be decomposed are those which are moving away from the electrode, and in this case when the intensity of the field is greatest the atoms are nearest together, so that the separating tendency which is the difference in the effects on the atoms is minimized as much as possible; while in the case of the negative electrode, when the tendency to produce a difference was greatest the distance between the molecules was greatest too, so that we see in this case the molecules will dissociate more easily at the negative than at the positive electrode. Again, we must remember that those molecules which are near to the positive electrode and moving away from it, must previously have been approaching the electrode, and that during this time the action of the electric field was to make the atoms come closer together. When the direction of motion of the molecule is reversed by reflection at the positive electrode, the action of the electric field in separating the atoms in the molecule is reversed, so that unless the course of the molecule is extraordinarily unsymmetrical it will be in the same state when it gets away from the electrode as it was before it approached it, and as it was not dissociated in the one case it will not be in the other.

Next let us suppose that the electric field is uniform, as in the experiments described above; then as there is no evidence for any considerable condensation of gas about the electrodes, we shall suppose that the density of the gas is approximately uniform. Since everything is uniform the molecules will dissociate most easily when they are moving for the longest time in the direction of the lines of force. Now according to the vortex atom theory of gases the vortex rings as they approach the planes which form the electrode will expand, and as they expand they move more and more slowly, so that the molecules will be moving for the longest time in the same direction in the neighbourhood of the electrodes. And just as in the non-uniform field the molecules will be more likely to dissociate at the negative electrode than at the positive, for at the positive electrode those molecules which are likely to be dissociated are those which are moving away from the electrode, but they must previously have been approaching it, during which time they were being pushed nearer and nearer together, so that at this electrode the molecules which have any tendency to be dissociated are those which have been specially prepared to resist this tendency, and as this is not the case at the negative electrodes the molecules will be dissociated most easily at this electrode.

Hence the conclusion we arrive at is, that whether the field be uniform or variable, dissociation is more likely to take place at the negative electrode than at the positive, and that the dissociation is more likely to take place close to the negative electrode than in the body of the gas; though if the field be very strong or the gas very weak the molecules in the body of the gas may get decomposed. Thus in the experiments described above, though in gases which are electrically strong, such as air and coal-gas, the discharge under certain circumstances could be confined to the neighbourhood of the negative electrode, yet in electrically weak gases, such as the vapours of turpentine and alcohol, the gas was under all circumstances (when the pressure was low) decomposed throughout the field, though the greater brightness of the layer near the negative electrode shewed that more gas was decomposed there than in other parts of the field.

There seems too in the case of the discharge through ordinary vacuum tubes considerable direct evidence that there is a considerable amount of decomposition going on near the negative pole, more so than in the rest of the field, for in the first place, the spectrum of the glowing gas surrounding the negative electrode generally shews lines, while the spectrum in the rest of the field is a band spectrum, and line spectra are believed to denote a simpler molecular constitution than band spectra: and secondly, the gas near the negative electrode is hotter than that in other parts of the field.

Let us trace some of the consequences of the gas being decomposed more easily at the negative than at the positive electrode. Since, according to our view, decomposition of the molecules means a spark, it follows that according to this theory a spark ought to pass more easily from a negative than from a positive pole, a result which was long since observed by Faraday (*Experimental Researches*, § 1501).

Again, decomposition, and therefore discharge, takes place the more easily the longer the molecules move continuously in the direction of the lines of force; thus the longer the average time

4

Gaseous Discharges: Further Developments (1891–1895)

Six months after their marriage on 2 January 1890, Thomson and his wife, Rose, moved from temporary accommodation at 15 Brookside, Cambridge, to a house owned by Caius College at 6 Scrope Terrace where they remained for nine years. Their new home became something of a social centre, with friends and staff from the Cavendish Laboratory being entertained, sometimes (for example, at the 'at home' held on Saturday afternoons) by Rose herself who seemed thoroughly to enjoy the company of scientists. Few of the visitors were ladies, the colleges and Cambridge in general being male-dominated societies. It was traditional for married fellows of Trinity College to dine in hall on Sunday nights and this Thomson did regularly.

Grotthus Chains and Faraday Tubes of Force

Around 1890, Thomson's research on gaseous discharges took a new direction in response to Hertz's recent work on electromagnetic waves, and to his own realisation that the evidence for discrete charges, especially in electrolysis, was overwhelming. His attempts to reconcile discrete charges with the Maxwellian tradition of a continuous electric medium led him to a model of 'Faraday tubes', which he envisaged as vortex lines and rings in the ether, which propagated electromagnetic effects and located charges. This new model was in conflict with the vortex model of the atom, which he abandoned, as for various reasons did a number of other physicists at around the same time.

The acquisition of a concept of discrete charge changed Thomson's interest in gaseous discharge. Prior to 1890, in accordance with Maxwell, he thought of discharge as relaxation of a strained state in the ether; a problem of energy dissipation. He now came to view discharge as a problem of charge transfer, and this is reflected in the changed nature of his experiments. Hertz's demonstration of electromagnetic waves in 1888, and his results, directed Thomson's attention to the velocity of discharge, which proved to be unexpectedly high and was difficult to interpret using charge transfer. Thomson's first use of aggregates of polarised molecules, the Grotthus chain, was devised largely to account for this unforeseen high discharge velocity.

Thomson began to speculate about chains in the gas in 1889. Hertz's experiments had shown a difference between the speed of propagation of electrical disturbances along a wire and through the surrounding air. According to Maxwell's theory the two would be the same. In a series of theoretical and experimental works Thomson accounted for the discrepancy, but he also exploited it to explain the striations observed in discharge. His developing ideas are illustrated in the sketch from his notebook reproduced in Figure 4.1. He suggested that striations represented standing waves, caused by interference between the electrical disturbance travelling along a conducting path through the gas and that travelling through the surrounding gas. For this he had to define the conducting path through the gas as localised, analogous to a wire, and different in some way from the surrounding gas. The sketch clearly shows the conducting path stretching, like a chain or wire, between the two electrodes. The electromagnetic disturbance in the surrounding gas is shown as loops or

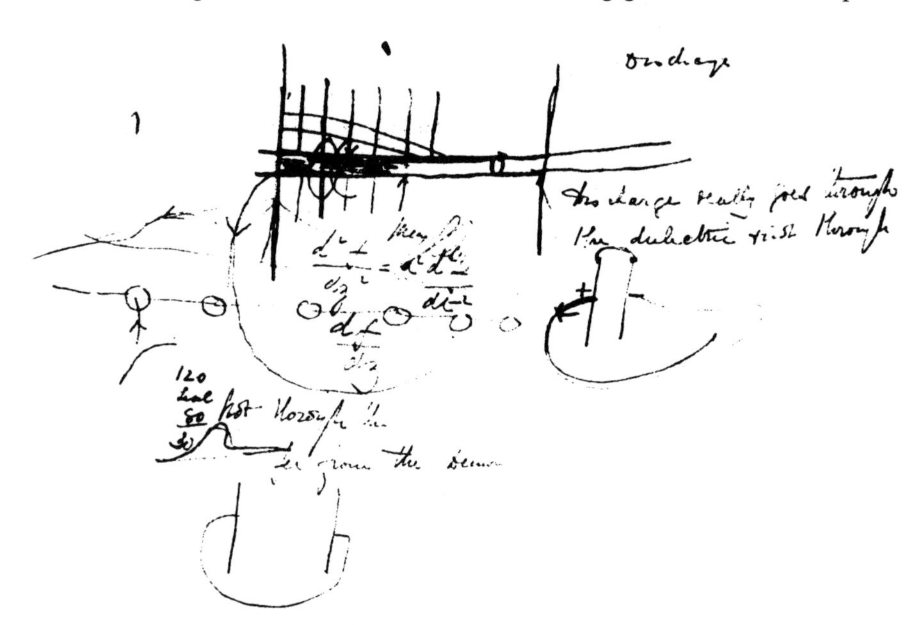

Figure 4.1 Sketch from Thomson's notebook illustrating the development of his ideas on gaseous discharges. The vertical lines represent striations which Thomson initially thought arose from interference between the passage of electricity through the main discharge path and the surrounding dielectric. Courtesy of Cambridge University Library [1]

circles around the 'chain'. Here Thomson was following his friend Poynting's recent ideas on the propagation of electromagnetic effects, and we see the beginnings of the Faraday tube theory discussed below. The vertical lines represent the nodes of the standing wave, i.e. the dark patches in the striations.

By the following year, this mechanism had changed and the chain idea developed, as Thomson realised that, 'There is another theory of molecular structure which is almost forced upon us by the laws of electrolysis and that is that the forces between atoms are electrical in their origin. On this theory atoms in the molecule are supposed to carry definite charges . . . ' [2]

Noting that the energy required to split up such a molecule must be large, he thought it unlikely that an electric field could accomplish this alone because 'collisions between molecules are probably so vigorous, that the electric forces trying to split up the molecules will have to commence their work afresh after each collision.' [3] He therefore sought ways in which the decomposition could be helped in some way and, to this end, proposed either the presence of an electric double sheet near the surface of one of the electrodes or the existence of short Grotthus chains.

The concept of chains, first introduced by Grotthus to explain electrolysis of liquids, involved molecules being polarized in the direction of the electric field with their positive ends facing towards the cathode (the negative electrode) and the negative ends towards the anode. Thomson suggested that in a gas subjected to an electric field, short chains of these polarized molecules were formed near the electrodes and that the splitting of the molecules passed 'from one molecule in the chain to another, being helped by the chemical forces between the molecules' [3]. This idea was developed with the aid of a new model of the electric field based on Faraday tubes.

About fifty years earlier, Michael Faraday had proposed the idea of tubes of electric force which threaded their way through space between positive and negative regions of electricity. Maxwell's theory of electromagnetism was based on this concept, but his heavy mathematical treatment tended to obscure any physical interpretation which might be assigned to them. Thomson ascribed reality to the tubes' existence by regarding them as structures composed of ether: entities that had both definite sizes and shapes. In addition to tubes which began and terminated on atoms, he proposed the existence of closed tubes which were present even in the absence of electric forces and which imparted 'a fibrous structure to the ether'. [4] The tubes, which he assumed were vortices, all had the same strength, the assumption being that when they terminated on a conductor, they had at their end a charge of electricity (positive or negative) equal in magnitude to that carried by a monovalent atom, such as hydrogen or chlorine, in electrolysis. Harking back to his fellowship thesis that all forms of energy were expressions of unseen motion, Thomson associated the *potential energy* of the electrostatic field to the *kinetic energy* of rotation of the ether inside and around the tubes. He argued that, within a molecule, short tubes connect the atoms and electrification is associated with chemical dissociation or rupture of these tubes.

Thomson's model for dissociation in the gaseous discharge, based on a combination of Grotthus chains and Faraday tubes of force, is illustrated in Figure 4.2. In the top sketch, molecules of the gas AB, CD, EF are shown polarised by induction due to the presence of a Faraday tube OP stretched between the electrodes. Attraction between the short tubes in the molecules and the neighbouring long tube occurs on account of their opposite polarities (rotations of vortices). This attractive force leads to bending and eventually to rupture of the tubes in the molecules and to the formation of new molecules in which one atom in a molecule becomes paired to one in a neighbour. The original tube starting out from P now terminates on atom F, creating in effect, a virtual anode. This same process can then occur at the next chain and so on, the disturbance thereby being transmitted from one end of the discharge to the other while the atoms have scarcely moved at all. The velocity of transmission may accordingly be very high, being given by the length of a chain divided by a molecular recombination time, which Thomson estimated to be about 10^{-11} seconds.

Thomson actually measured the velocity of transmission of a gaseous discharge experimentally (using a discharge tube 50 feet long and a rotating mirror method [5]) and found it to be comparable to the speed of light. The result, which would not have been a problem on his pre-1890 view of the discharge being a release of energy from a strained ether, was inexplicable in terms of charged atoms travelling at such a high velocity since the applied voltage was insufficient to provide the required kinetic energy. The Grotthus chain model provided a satisfactory alternative explanation. It also provided a new explanation for the striations, as the discharge was broken up into a series of separate chains or pieces.

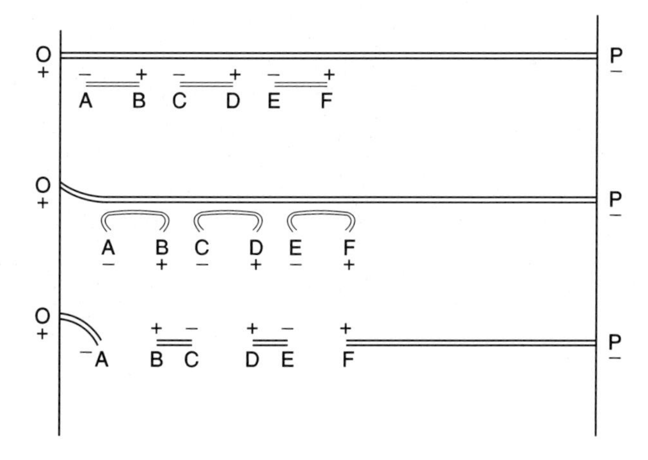

Figure 4.2 Faraday tubes of force and their interaction in the Grotthus chain model of the gaseous discharge

Discharges in bell jars

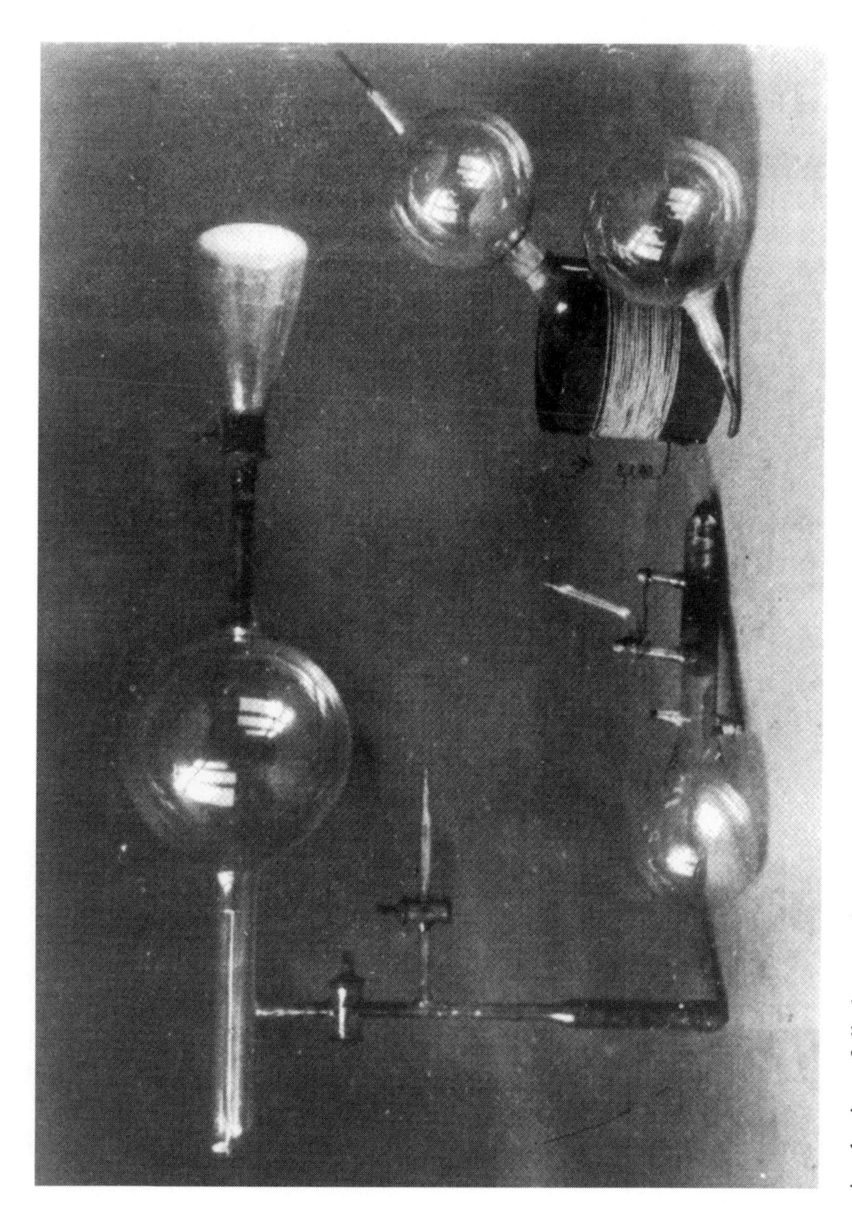

A selection of discharge tubes

Electrodeless Discharges

A continuing source of concern in experiments on gaseous discharges was the effect of the electrodes. The electric breakdown strength of a gas varied with the shape, material and distance apart of the electrodes. Furthermore all discharges were asymmetrical, the spatial disposition of the emitted light, in particular, distinguishing anode from cathode. This last feature was puzzling and clearly hard to explain on a model of dissociation that split a molecule into essentially equivalent positively and negatively charged atoms.

In 1891 Thomson conducted experiments on discharges excited by an oscillatory current flowing in a coil situated on the *outside* of the evacuated vessel, thereby eliminating electrodes and offering the possibility of identifying phenomena solely associated with the gas itself. [6] Here, again, he owed a debt to Hertz, whose new high-frequency techniques made these experiments possible.

6 Scrope Terrace, Cambridge
April 18th 1891

Dear Threlfall
I have been making a good many experiments lately without electrodes of any kind internal or external. If you wish to avoid electrodes, I think the method might be worth trying.

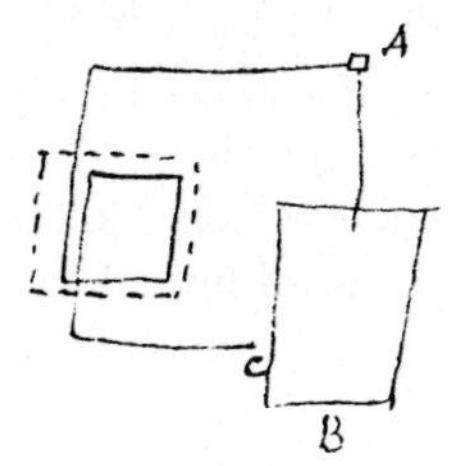

AB is a Leyden jar, A connected to Earth, this is connected up to Wimshurst Machine and the jar discharges by sparking across an air gap at C. The continuous line connecting AC is either a wire or what I find more easily insulated a glass tube of smallish bore filled with mercury. The oscillatory charge of the jar produces very rapidly alternating currents in this circuit (which corresponds to the primary of an induction coil) and there is consequently a field of very great electromotive force surrounding the circuit. This is quite enough to produce a discharge in the exhausted tube represented by the dotted line in the figure the discharge flows round in a continuous circuit going through [the] gas the whole of the way. I have made a large number of experiments with this arrangement and got some very interesting results. . . [7]

A sketch of an arrangement used by Thomson for producing electrodeless discharges in a spherical bulb is illustrated in Figure 4.3. Under the right conditions the discharge formed an intense circular thread of light situated

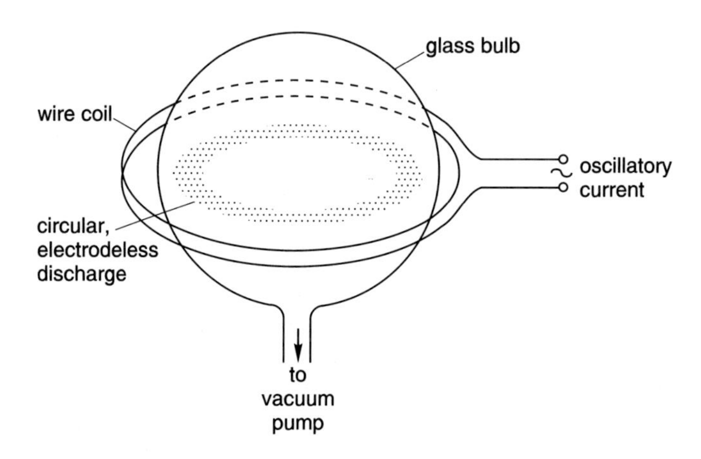

Figure 4.3 The electrodeless discharge

near the walls of the tube. No striations were observed, a feature which Thomson explained by the rapid ($\sim 10^6$ Hz) switching of the field, resulting in the superposition of two discharges in opposite directions, 'the places of maximum luminosity in one fitting into those of minimum luminosity in the other' [7]. The two discharges could be separated by a magnetic field applied at right angles to the plane of the ring. He also found that such a transverse field inhibited the initiation of the discharge, while one lying in the plane of the discharge facilitated its passage. These effects Thomson attributed to Grotthus chains of molecules being directed out of line of the preferred direction of the discharge for a transverse magnetic field but brought together in the longitudinal case. With oxygen in the bulb, Thomson observed a phosphorescent glow which lasted for several minutes after cessation of the discharge. For other gases, he detected spectral lines in the emission and drew attention to the advantages of electrodeless discharges for spectroscopic purposes.

6 Scrope Terrace
25 February 1892

Dear Threlfall
I am much ashamed of my laziness as a correspondent but we have all been very much upset for some time by the illness and death of my wife's father Sir George Paget . . .

. . . I think you will like the discharge without electrodes when you try them. I have got them so bright that I think it is not at all impossible that they may in time give us a means of illumination far superior to incandescent lamps.

. . . I have been busy seeing another edition of Maxwell through the press and writing a volume of notes . . . [8]

Although elimination of the electrodes offered promise for a better understanding of gaseous discharges, their oscillatory nature added complications with which Thomson's theories could not adequately cope. Nevertheless, experiments in which the pressure in the bulb was varied did provide evidence for his belief that a vacuum was a perfect insulator and also demonstrated that, at low pressures, gases were surprisingly good conductors. Thomson concluded that most of the resistance in the direct-current discharge lay in the difficulty of transferring electricity from the electrodes to the gas. He returned to further investigations of electrodeless discharges in the 1920s. [9]

In 1893 Thomson formed the Cavendish Physical Society – not really a society in the normal sense of the word, but an informal gathering which met fortnightly on Tuesdays during University terms. At these meetings a member of the teaching staff, sometimes Thomson himself, a research student or perhaps a visitor from another department, gave a presentation which was followed by an open discussion. Topics included the speaker's own work or recent research discoveries reported in journals, English or foreign. Sometimes the discussion was apt to flag, in which case Thomson as Chairman kept it alive. When a new discovery was to be presented, the room could be full to overflowing. Around 1895, Rose suggested that the meetings became more sociable occasions and began the tradition of serving tea before the presentation. Every research student was asked to talk about his work at least once during his time in the laboratory and although, for some at least, this must have been an intimidating experience, it provided them with valuable experience in speaking publicly and in debating.

When not lecturing, Thomson worked mostly at home in the mornings. On these days he would generally not arrive at the Cavendish until one o'clock, whereupon he would walk round the laboratory discussing work with any of the research students who had decided that such an encounter was preferable to lunch! Thomson's afternoons were spent on experimental work, with the assistance of laboratory attendants on whom he relied a great deal. Most apparatus used by Thomson and other researchers was made in the laboratory and the skills of the workshop staff were therefore exceedingly important regarding the success or otherwise of a particular investigation.

F.W. Aston, who was to achieve fame for development of the mass spectrograph bearing his name, wrote in *The Times* after Thomson's death:

> Working under him never lacked thrills. When results were coming out well, his boundless, indeed childlike, enthusiasm was contagious and occasionally embarrassing. Negatives just developed had actually to be hidden away for fear he would handle them while they were still wet. Yet, when hitches occurred, and the exasperating vagaries of an apparatus had reduced the man who had designed, built and worked with it to baffled despair, along would shuffle this remarkable being, who, after cogitating in a characteristic attitude over his funny old desk in the corner, and jotting down a few figures and formulae in his tiny, tidy handwriting, on the back of somebody's fellowship thesis, or an old envelope, or even the

> laboratory cheque book, would produce a luminous suggestion, like a rabbit out of a hat, not only revealing the cause of the trouble, but also the means of a cure. This intuitive ability to comprehend the inner working of intricate apparatus without the trouble of handling it appeared to me then, and still appears to me now, as something verging on the miraculous, the hallmark of a great genius. [10]

Electrolysis of Steam and Condensation

In 1893 Thomson published *Notes on Recent Researches in Electricity and Magnetism*, a book intended as a supplementary volume to Maxwell's great work *Electricity and Magnetism*, the second edition of which Thomson had just edited. Like Maxwell's treatise, Thomson's book makes for difficult reading and is certainly not a student textbook. Both authors had graduated in the Mathematical Tripos at Cambridge which taught and demanded formidable mathematical skills and these are evident in *Recent Researches*. An exception, however, is Chapter 2 in which he summarized, for the first time in English, the extensive literature then existing on discharge in gases and provided a *qualitative* explanation for many of the observed phenomena based on Faraday tubes and Grotthus chains. For his explanation of discharge phenomena, the most important part of this theory was that Faraday tubes could terminate only on atoms. Here a defined electric charge was located, as a boundary effect between the end of the Faraday tube and the atom. In a molecule, two oppositely charged atoms were linked by a Faraday tube, and the molecule was neutral. To carry an electric charge the atom had to be wrenched free from the molecule. Thus electrification required free atoms and was always accompanied by chemical dissociation.

Over the next few years Thomson increasingly came to view changes in electrification and chemical action as synonymous, postulating a variety of different molecular aggregates and electric double layers to explain the mechanisms involved. By 1895 this work had led him to an increased understanding of the relation of charges and atoms, and a new theory of their interaction expressed in his paper on the gyroscopic atom (discussed below).

His initial experiments were on the electrolysis of steam. It was apparently during the writing of *Recent Researches* that Thomson first learned of experiments conducted by Perrot on the electrolysis of steam. [11] Since these seemed to provide the first direct evidence for chemical dissociation being an essential accompaniment of discharge – a vital ingredient of Thomson's world – he decided to repeat the experiments. The experimental arrangement he designed (Figure 4.4) was a modification of that used by Perrot. A discharge passing between the electrodes A and B electrolysed steam generated from heated water in the vessel H. The liberated gases (oxygen and hydrogen) were collected over water. When the electrodes were close, i.e. the sparks short, Thomson found a most surprising result, namely that hydrogen was evolved at the *positive* elec-

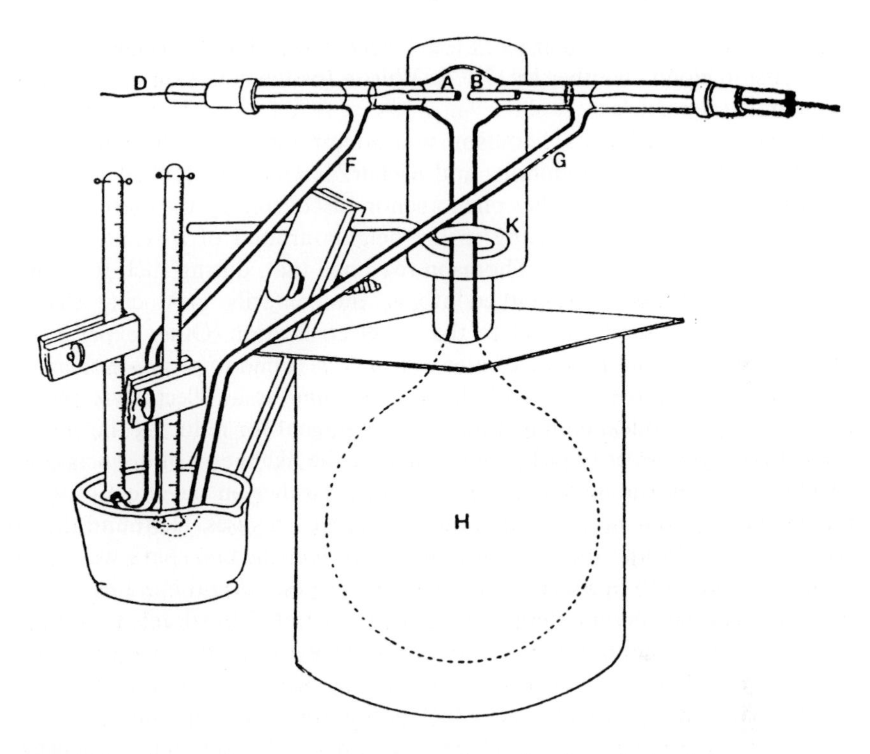

Figure 4.4 Thomson's apparatus for research on the electrolysis of steam

trode, in direct contradiction to Perrot's findings and to the situation for the electrolysis of water in which hydrogen is released as a positive ion at the negative electrode. The observation threatened to undermine his whole theory of gaseous discharge and its analogy with electrolysis. Before publication of the results [12], however, Thomson experimented with longer sparks and happily found the opposite, and expected, effect that hydrogen was liberated at the anode. He explained the contrary effect using short sparks in terms of the steam acting as if it were a mixture of pure hydrogen and oxygen, permitting the hydrogen to acquire a negative charge and oxygen a positive one. Although somewhat *ad hoc*, this explanation found support in experiments on discharges through pure hydrogen or oxygen alone, which revealed that the acquired charges were as proposed.

Later in the same year, Thomson investigated the enhancement of condensation in a jet of steam when 'an electrode from which electricity is escaping is placed close to the origin of the jet', typically a nozzle. [13] He suggested that the intense electric field around charged atoms of the gas carrying the electricity led to a reduction in the potential energy associated with the surface tension of the water drop. Even for dust-laden air, in which nuclei facilitate condensation, he found that electrification could enhance drop formation by diminishing their tendency to evaporate. This result turned out to have particular practical significance in Thomson's later work (see Chapter 5) and in the development of the expansion cloud chamber by C.T.R. Wilson for the detection of the tracks of fundamental particles.

The experimental results Thomson was attempting to explain had in fact been reported earlier by Helmholtz and Richarz in Germany [14] as well as by other English scientists. Another phenomenon discovered by the Germans was that chemical reactions occurring in the neighbourhood of a steam jet also produced dense condensation. Thomson reasoned that, during such reactions, molecules dissociated into charged atoms, momentarily producing electric fields which had a similar effect to that of electrification. Other experiments of a converse character, in which water vapour was found to promote chemical action, he interpreted in terms of the water acting as an electrolyte (which would promote chemical combination) or as an agent for reducing the electric force between molecules (which would enhance the likelihood of dissociation).

These results prompted Thomson to research further on the effect of water vapour on the voltage required to initiate discharges in gases. He found that in the case of very dry hydrogen, the voltage to produce the first spark was about twice that required to produce subsequent discharges. Several minutes of resting were needed to permit the gas to recover its initial breakdown strength. Furthermore, he found that the first spark was always brighter and appeared to start 'with a rush'. A similar but much reduced effect was found with damp hydrogen, the voltage to initiate the first breakdown being only about 10 per cent higher than for following discharges. Thomson drew an analogy between these results and the supercooling of a vapour below the boiling point or that of a liquid below its normal freezing point in the absence of foreign substances which act as nuclei. In the case of the discharge, Thomson argued that the passage of a spark 'is preceded by the condensation of some of its molecules into a more complex state of aggregation'. [13] Once these are formed, i.e. after the first discharge, the voltage needed is less – indeed even the initial spark can be maintained with a lower voltage than that required to initiate it. The formation of these aggregates is facilitated by the presence of water vapour (by the electrolytic atom referred to above) and so the breakdown voltage does not vary greatly between the first and subsequent discharges in the case of damp hydrogen. 'On this view', Thomson concluded 'the discharge through a gas does not consist of tearing the atoms of a single molecule apart, but rather in tearing atoms from off a complex aggregate of molecules.' [13] It would appear that, at this juncture, Thomson was beginning to abandon the concept of linear Grotthus chains in favour of more compact globular-like aggregates.

Electricity of Drops

In 1894 Thomson furthered his views on the connection between electrification and chemical action by undertaking a series of investigations on the charge acquired by falling drops of liquid. [14] In these he was stimulated by experiments of Lenard [15] and Kelvin and Maclean [16] who had shown that when drops of distilled water fall on a plate wetted by the water, they acquire a positive charge, while the surrounding air becomes negatively electrified.

The approach Thomson used was similar to that of Lenard's (see Figure 4.5). Water drops falling through air, water vapour and hydrogen acquired, respectively, positive, zero and negative charges. Experiments with other liquids and other gases showed varying degrees of electrification. The observation that a drop falling through a gas of its own kind produced no charging led Thomson to conclude that 'the electrification owes its origin to chemical processes' and that 'over the surface of the drop a substance is formed which is in a state intermediate between that of complete chemical combination and complete separation . . . in which the connexion between the constituents is so loose that they can easily be shaken apart.' [14]

Thomson extended this model of partial chemical combination at liquid surfaces – in effect a loosely bound double layer of electrification – to solids, suggesting that electricity developed by friction and also the charge acquired by metals when irradiated with ultraviolet light could be explained by separation of the layers. He also applied the model to discharge phenomena. In particular he ascribed the fall of potential near the cathode to the difficulty that positive electricity had in passing from the gas to the cathode. The formation of a chemical compound between the gas and the electrode might ease the transfer of electricity between the two, and for this to happen the gas and the metal must have the correct relative charges or there must be some mechanism for interchanging charges. Implicit here is Thomson's explanation for the asymmetry between anode and cathode: at the anode the gas and the metal must have the correct relative charges for chemical combination; at the cathode they have the wrong ones.

The Gyroscopic Atom and Chemical Combinations

The liquid drop experiments had suggested to Thomson that the energy of an atom when charged with a unit of positive electricity is not the same as when it is charged in the opposite sense. Certain atoms (for example, oxygen and chlorine) seem to have a preference for acquiring a negative charge, while others (for example, hydrogen) prefer a positive one. In two papers published in 1895, Thomson examined and pursued the question of the preferred charge carried by a given atom, the circumstances in which it may carry the opposite charge, and the necessity of charge exchange in certain chemical reactions. [17,18] In electrolysis of liquids, the preferred or normal charge invariably occurs but, as he had found during his experiments on the electrolysis of steam for example, the situation for gases is much more variable. Thomson's understanding of the preference of a particular atom for a certain sign of charge rested on an ingenious dynamical model of an atom which he pictured as consisting of a number of gyrostats all spinning in the same sense around

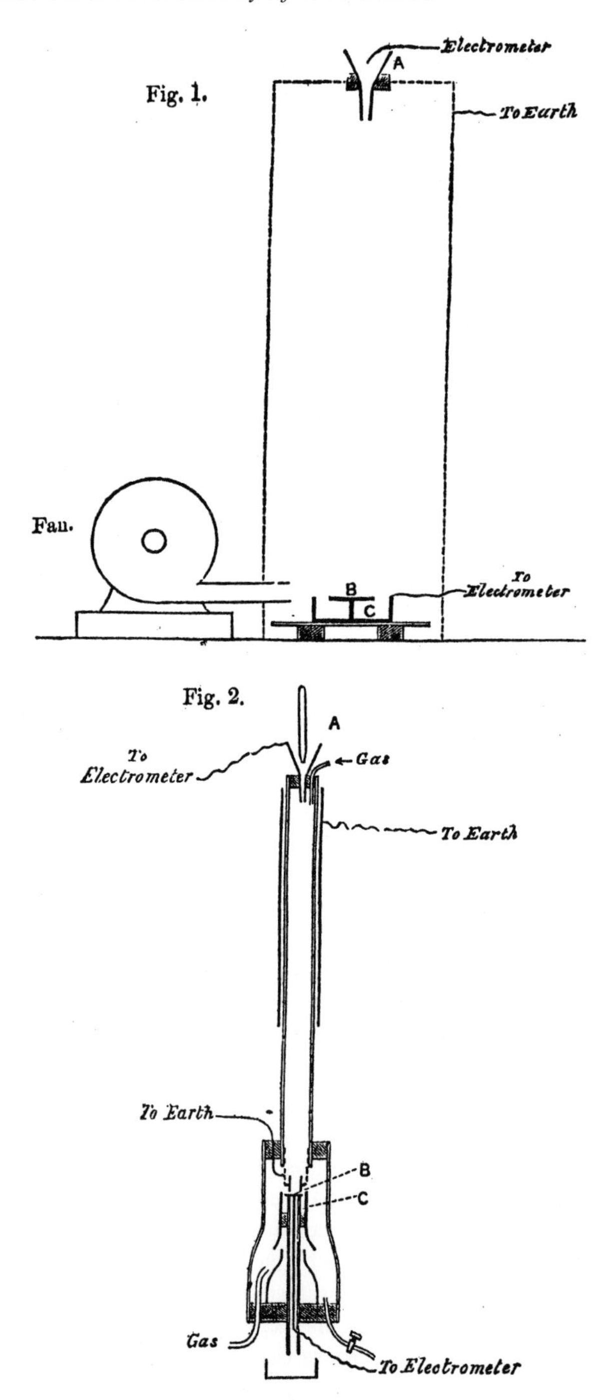

Figure 4.5 Apparatus for investigations of the electrification acquired by falling drops

outwardly facing normals to its spherical surface. This model of a *gyrostatic atom* introduced a different type of structure to that of his previous vortex atom, while at the same time incorporating Thomson's other concept of a discrete charge at the end of rotating Faraday tubes of force. Relevant parts of the paper published in the *Philosophical Magazine* are reproduced on p. 94.

With reference to Figure 4.6, the dynamical properties of the model can be summarised as follows. The flywheel CD, mounted on gymbals, is set spinning with its axis vertical and the remainder of the system stationary. Consider now rotation of the framework around the axis AB. If this rotation is in the same sense as that of the flywheel, the whole system rotates steadily. On the other hand, if the rotations are in opposite senses, the axis of the wheel at first wobbles, then, as the disturbance becomes more violent, the flywheel eventually turns turtle. Now with the wheel rotating in the same direction as the framework, the whole system rotates smoothly as before. That the second case costs more energy to execute than the first becomes evident if, rather than being unrestrained, the flywheel is imagined to be held in its original position by springs. Then, instead of toppling as it did in the second case, the wheel would take up a position with its axis inclined to the vertical with potential energy stored in the springs.

Applying the model to, say, a hydrogen atom, Thomson assumed that the acquisition of a positive charge (corresponding to a rotating vortex Faraday tube *leaving* the atom) was equivalent to adding rotations in the same sense, with no change in the potential energy – a favourable situation. For the case of real atoms such a situation may, he surmised, even lead to a decrease in the potential energy. In contrast, the addition of a negative charge (corresponding to a Faraday tube *arriving* at the atom) is equivalent to the case of counter-rotating bodies and leads to an increase in the potential energy. For chlorine, an electronegative element, Thomson assumed the gyrostats to be rotating in the opposite sense to that of hydrogen, hence favouring acquisition of a negative charge by this element. Furthermore, on this model, the force of attraction between atoms carrying their preferred charges, e.g. H^+ and Cl^-, is expected to be greater than that between the same atoms with opposite charges, i.e.

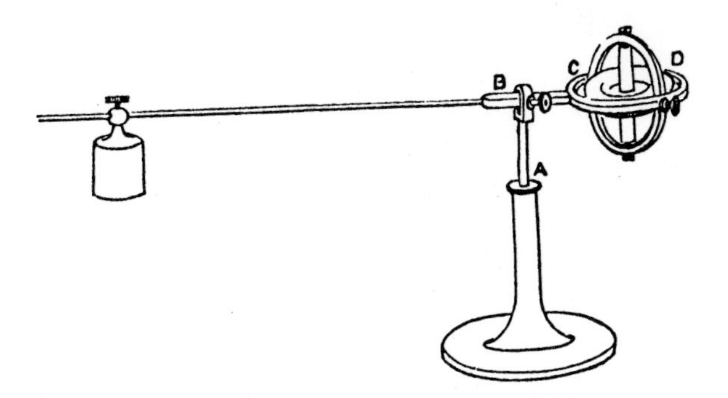

Figure 4.6 Gyroscopic arrangement for modelling the atom [18]

between H^- and Cl^+. This follows because, in the latter case, as the ions are brought together there will be a decrease in the potential energy due to the normal electrostatic effect but an increase due to the effect on the gyrostats of the rotating Faraday tube. Indeed, at small distances, Thomson surmised that the force of attraction could become one of repulsion.

In considering a specific example of chemical activity, namely combination of the gaseous molecules hydrogen (H_2) and chlorine (Cl_2) to form hydrochloric acid (HCl), Thomson drew on the above ideas and introduced a specific type of aggregrate – the ring-like structure shown in Figure 4.7. Such a disposition of molecules minimises the cost in energy of the first step in which negatively charged hydrogen atoms and positively charged chlorine atoms on neighbouring molecules exchange their charges prior to combination.

Summary

By 1895, Thomson's theory of discharge, although superficially similar to that with which he had started out in 1883, had changed radically in concept. He maintained throughout that gaseous discharge was similar to electrolysis, both processes requiring chemical dissociation. But his original interest in the dissipation of energy during discharge had changed to a realisation of the importance of charge separation and transfer. The legacy of his early background is still evident in 1895: hydrodynamical analogies for the Faraday tubes, and the idea that the energy of the electric field is the kinetic energy of an unseen mechanism, the Faraday tube.

The years 1890–1895 saw Thomson focus increasingly on the importance of electric charges to his fundamental concern of the relation of ether and matter.

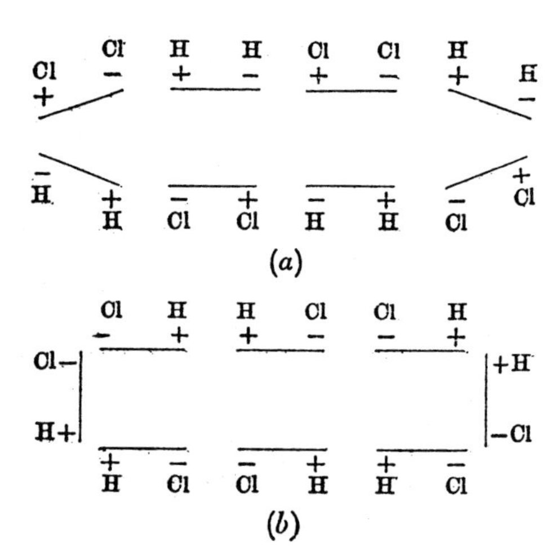

Figure 4.7 Dissociation of hydrogen and chlorine molecules in a ring-like aggregate to form hydrochloric acid

By 1896, he was writing: ' . . . the relation between matter and electricity is indeed one of the most important problems in the whole range of physics . . . These relations I speak of are between *charges* of electricity and matter. The idea of charge need not arise, in fact does not arise as long as we deal with the ether alone' (italics, our emphasis). [19]

By 1895, Thomson had a clear mental imagery of the nature of an electric charge that was necessarily related to the chemical nature of the atom, and this placed him in a unique position among physicists.

At the same time, he had acquired a great deal of knowledge and understanding of discharge experiments, of the way apparatus behaved, of what to do to bring certain effects to the fore, and of when a discharge would pass, which was of vital importance to him over the next few years. He was also, however, by 1895, fairly committed to the view that the discharge mechanism involved aggregates of ions, larger than an atom. This to some extent mitigated against his realisation of the nature of X-ray induced conductivity in 1896 and of cathode rays in 1897.

References

Cambridge University Library holds an important collection of Thomson manuscripts, classmark ADD 7654, referred to here as CUL ADD 7654, followed by the particular manuscript number. Other archives, and a more complete bibliography, may be found in: Falconer, I., Theory and Experiment in J.J. Thomson's work on Gaseous Discharge, PhD Thesis, 1985, University of Bath.

[1] CUL ADD 7654 NB35a.
[2] CUL ADD 7654 NB35a f22.
[3] THOMSON, J.J., *Philosophical Magazine*, **29** (1890) 358.
[4] *Notes on Recent Researches in Electricity and Magnetism*, 1893, Oxford: Clarendon, 2.
[5] THOMSON, J.J., *Philosophical Magazine*, **30** (1890) 129.
[6] THOMSON, J.J., *Philosophical Magazine*, **32** (1891) 321, 445.
[7] CUL ADD 7654 T26.
[8] CUL ADD 7654 T27.
[9] THOMSON, J.J., *Philosophical Magazine*, **4** (1927) 1128.
[10] ASTON, F., Obituary of Thomson. *The Times*, 4 September 1940.
[11] PERROT, A., *Annalen der Physik*, **61** (1861) 161.
[12] THOMSON, J.J., *Proceedings of the Royal Society*, **53** (1893) 90.
[13] THOMSON, J.J., *Philosophical Magazine*, **36** (1893) 313.
[14] THOMSON, J.J., *Philosophical Magazine*, **37** (1894) 341.
[15] LENARD, P., *Annalen der Physik*, **46** (1892) 584.
[16] LORD KELVIN AND MACLEAN, M., *Philosophical Magazine*, **38** (1894) 225.
[17] THOMSON, J.J., *Proceedings of the Royal Society*, **58** (1895) 244.
[18] THOMSON, J.J., *Philosophical Magazine*, **40** (1895) 511.
[19] CUL ADD 7654 NB40.

value of the integral in (4) or

$$\int_0^{2\cdot 40483} x f(x)\, J_0(x)\,.dx = 0\cdot 6117,$$

and hence by (4), as $J_1(2\cdot 40483) = 0\cdot 519$,

$$A_1 = \frac{2 \times 0\cdot 6117}{(0\cdot 519)^2 (2\cdot 405)^2} = 0\cdot 7852.$$

To find A_2. The 25 equidistant ordinates of the original curve (straight line) are the ordinates of points in the curve MBLBBP (fig. 2), the abscissæ being taken from Table II. We found it convenient to multiply the abscissæ by 5 and represent in inches. Our ordinates were multiplied by 10 and represented in inches. The actual area was 2·36 square inches and one-fiftieth of this, or 0·0472, is the value of the integral in (4).

Hence, as $J_1(5\cdot 5201) = \cdot 3403$,

$$A_2 = \frac{2 \times 0\cdot 0472}{(0\cdot 3403)^2 \times (5\cdot 5201)^2} = 0\cdot 0268.$$

We need not show the curves used in finding A_3 and A_4. The area of fig. 2 is the positive area MLNOM minus the area LQPNL, but one need not think about whether an area is positive or negative. It is only necessary to start the planimeter-tracer from the point numbered 0 in every case, and go from 0 to M and along the curve in the direction of the increasing numbers to 24, then along the axes of abscissæ, ending at the point 0 from which we started.

XLVIII. *The Relation between the Atom and the Charge of Electricity carried by it.* *By* J. J. Thomson, *M.A., F.R.S., Professor of Experimental Physics, Cambridge**.

IN the electrolysis of solutions, the persistency of the sign of the electric charge carried by an ion is almost as marked a feature as the constancy of the magnitude of the charge. Thus the hydrogen ion always carries a positive charge, the chlorine ion a negative one. In the electrolysis of gases, however, the sign of the charges carried by the atoms of the different elements is much more variable: here an atom of hydrogen does not always carry a positive charge, nor an atom of chlorine always a negative one; each of these

* Communicated by the Author.

2 N 2

atoms sometimes carries a negative charge, sometimes a positive one. But though in the gaseous state the atoms do not restrict themselves to charges of one sign, there are many phenomena which prove that even in this state the atoms of the different elements behave differently with respect to positive and negative charges. v. Helmholtz explained this behaviour of the elements by supposing that there is a specific attraction between the atom and its charge; that the zinc atom, for example, attracts a positive charge more powerfully than it does a negative one, while an atom of chlorine, on the other hand, attracts a negative charge more powerfully than it does a positive one.

The connexion between ordinary matter and the electrical charges on the atom is evidently a matter of fundamental importance, and one which must be closely related to a good many of the most important chemical as well as electrical phenomena. In fact a complete explanation of this connexion would probably go a long way towards establishing a theory of the constitution of matter as well as of the mechanism of the electric field. It seems therefore to be of interest to look on this question from as many points of view as possible, and to consider the consequences which might be expected to follow from any method of explaining, or rather illustrating, the preference which some elements show for one kind of electricity rather than the other.

The following method of regarding the electric field seems to indicate that this effect may be illustrated by simple dynamical appliances. The charges on the atoms are regarded as the ends of Faraday tubes (see J. J. Thomson, 'Recent Researches in Electricity and Magnetism,' p. 2): each unit of positive charge is the origin, each unit of negative charge the termination of such a tube. The magnitude of the unit charge is here taken equal to the charge carried by a monovalent ion. When these tubes spread into the medium they give rise to Maxwell's *Electric Displacement*, and the motion of the tubes through the medium produces a magnetic field.

Now suppose that the medium forming these tubes possesses rotatory motion: we may imagine, for example, that the tubes are bundles of vortex filaments, the axis of rotation being parallel to the axis of the tube. The total amount of vorticity which starts from any solid totally immersed in a liquid is zero; to satisfy this condition, we may suppose that there is slipping between the walls of the bundle of vortex filaments and the surrounding liquid, or, what amounts to the same thing, that there is on the surface of the bundle a film of negative vorticity whose strength is equal and opposite to the positive vorticity of the core. To fix our ideas, let us suppose that the rotation in the core is related to the direction of the axis of the tube (the line running from the origin to the end of the tube) like rotation and translation in a right-handed screw.

Now let us consider the atoms on which these tubes end. Let us suppose that these atoms have a structure possessing similar properties to those which the atoms would possess if they contained a number of gyrostats all spinning in one way round the outwardly drawn normals to their surface. Then one of these atoms will be differently affected by a Faraday tube, and will possess different amounts of energy according as the tube begins or ends on its surface. For if, when the tube starts from the atom, the rotation in the core of the tube is in the same direction as the rotation of the gyrostats, then when a tube ends on the atom the rotation in the tube will be in the opposite direction to that of the gyrostats. Thus in the first case the tube will tend to twist a gyrostat in the same direction as that in which it is already spinning, while in the second case it will tend to twist it in the opposite direction. Now if we try to rotate a gyroscopic system, its behaviour when we try to make it rotate in the direction in which the gyroscopes are spinning is quite different from its behaviour when we try to spin it in the opposite direction.

Thus let fig. 1 represent a balanced gyrostat which can

Fig. 1.

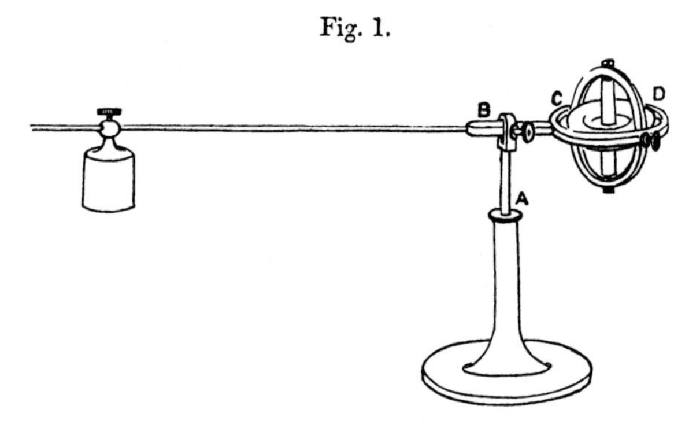

rotate as a whole about the vertical axis A B ; the heavy fly-wheel C D is supported so that its axis can move freely in a vertical plane. Let this fly-wheel be in rapid rotation with its axis vertical ; the framework of the gyrostat will not tend to rotate. Now if we set the framework rotating about the vertical axis, the behaviour of the system, if we make the framework rotate in the direction in which the fly-wheel is spinning, will be very different from its behaviour when we make the framework rotate in the opposite direction. If the framework rotates in the same direction as the fly-wheel, the axis of the wheel remains vertical, and the whole system goes on rotating quietly until stopped by friction. If, however, we attempt to set the framework rotating in the opposite direction to the fly-wheel, the axis of the wheel begins to wobble about, the disturbance gets more violent until finally the fly-wheel topples right over; then the fly-wheel is rotating in the same direction as the framework, and the rotation goes smoothly on. If the axis of the fly-wheel were held in its original position by springs, then in the first case these springs would not be stretched; but in the second case they would, and the axis of the fly-wheel would take up a position inclined to the vertical, the angle it made with the vertical depending on the stiffness of the springs and the moment of momentum of the fly-wheel. Thus in the second case the attempt to make the framework of the gyrostat rotate would be accompanied by the storing up of potential energy in the system due to the stretching of the springs, while there would be no such storage of potential energy in the first case.

Suppose, now, that an atom of hydrogen possesses a structure analogous to this gyroscopic system with springs, the gyrostats rotating in the same direction as the fluid in a Faraday tube *leaving* the atom. Then, since a charge of negative electricity on the atom implies the *incidence* of a tube on the atom, when the hydrogen atom is charged negatively the rotation in the tube connected with the atom is in the opposite direction to that of its gyrostats. The negative charge will thus cause an increase in the potential energy of the atom, whereas a positive charge when the rotation in the tube is in the same direction as that of the gyrostats does not cause any such increase. Thus the internal potential energy of the hydrogen atom will *cæteris paribus* be greater when it has a negative charge than when it has a positive one. In the case of a strongly electronegative element such as chlorine, we suppose that the gyrostats in the atom are rotating in the opposite direction to those in the hydrogen atom, *i. e.* that in the chlorine atom the rotation in the gyrostats is in the *opposite* direction to that of the liquid in a Faraday tube *leaving* the atom: thus the chlorine atom will, *cæteris paribus*, have more internal potential energy when charged with positive electricity than it has when charged with negative.

The existence of the property conferred by these gyrostats

would call into play forces between two charged atoms placed very close together, in addition to those given by the ordinary laws of electrostatics ; it would make, for example, the attraction between a negatively charged hydrogen atom and a positively charged chlorine one less than that between a positive hydrogen atom and a negative chlorine one at the same distance apart. For imagine, in the first case, the atoms to approach a little closer together, then, besides the diminution in the potential energy due to the ordinary electric forces between the atoms, there will be an *increase* in the potential energy from the increase in the effect on the gyrostats due to the rotation in the Faraday tubes ; while in the second case, when the hydrogen is positive and the chlorine negative, this increase will not take place. Thus the diminution in the potential energy due to a given diminution in the distance between the atoms is less in the first case than in the second, and consequently the attraction between them is smaller in the first case than in the second. If we could reach a place where, as the distance between ths atoms diminished, the increase in the potential energy due to the effect of the gyrostats was numerically greater than the diminution in the potential energy due to the electrostatic attraction, then the oppositely-charged atoms would repel instead of attracting each other.

Hydrodynamical Illustration.

The following illustration also indicates that the force between two electric charges may be modified by the electrochemical properties of the atoms carrying the charges.

In a cylindrical column of rotating fluid the pressure increases with the distance from the axis of rotation, so that the average pressure over a cross section of the cylinder is less than the pressure at the surface of the cylinder. When a solid is immersed in a liquid where the pressure is uniform, the pressures of the liquid on the solid form a system of forces in equilibrium. Now suppose that a column of the liquid abutting on the solid acquires rotation, the pressure on the part of the solid in contact with the column will be less than the pressure outside, the pressures on the solid will no longer be in equilibrium. The defect in pressure over the cross section of the column will give rise to a tension acting on the solid. This tension is equal to the excess of the pressure over the cross section, when the pressure is uniform and equal to that at the surface of the cylinder, over that actually exerted over the area by the rotating liquid. If the rotating column is a cylinder containing a number of vortex filaments mixed up with irrotationally moving liquid, the pressure over a cross section of the cylinder will depend upon the distribution of the vortex filaments in the cylinder. Let the cylinder be a right circular one. Let v be the velocity, and p the pressure at a distance r from its axis, ρ the density of the liquid; then we have

$$\rho \frac{v^2}{r} = \frac{dp}{dr}. \quad \ldots \ldots \quad (1)$$

The pressure over the cross section of the cylinder is equal to

$$\int_0^b 2\pi r p \, dr,$$

where b is the radius of the cylinder. Integrating this by parts, we find that the pressure Π over the cross section is given by the equation

$$\Pi = P\pi b^2 - \int_0^b \pi r^2 \frac{dp}{dr} dr$$

$$= P\pi b^2 - \int_0^b \tfrac{1}{2}\rho v^2 2\pi r \, dr \quad \ldots \quad (2)$$

by equation (1), P is the pressure at the surface of the cylinder

The tension Δ exerted by the cylinder on the solid is given by the equation

$$\Delta = P\pi b^2 - \Pi$$

$$= \int_0^b \tfrac{1}{2}\rho v^2 2\pi r dr. \quad \ldots \ldots \quad (3)$$

= kinetic energy per unit length of the cylinder. (4)

Since

$$2\pi r v = \text{vorticity inside a circle of radius } r,$$

we can, if we know the distribution of vorticity, easily calculate by means of equation (3) the value of Δ.

Let us suppose that if all the vortex filaments were collected round the axis to the exclusion of the irrotationally moving liquid, they would occupy a cylinder of radius a. Let ζ be the rotation in the vortex filament, and let

$$\int_0^b 2\pi r \zeta dr = m.$$

Then we find that when the vortex filaments are as close to the axis of the cylinder as possible,

$$\Delta = \frac{\rho m^2}{16\pi} + \frac{\rho m^2}{4\pi} \log \frac{b}{a}.$$

When the vorticity is uniformly distributed over the cross section of the cylinder,

$$\Delta = \frac{\rho m^2}{16\pi}.$$

When the vorticity is all as near to the surface of the cylinder as possible,

$$\Delta = \frac{\rho m^2}{16\pi}\left\{1 - 2\frac{(b^2-a^2)}{a^2}\left(1 - \frac{(b^2-a^2)}{a^2}\log\frac{b^2}{b^2-a^2}\right)\right\}.$$

We may expand the right-hand side of the last equation, and get

$$\Delta = \frac{\rho m^2}{8\pi}\left\{\frac{1}{3}\frac{a^2}{b^2-a^2} - \frac{1}{4}\frac{a^4}{(b^2-a^2)^2} + \frac{1}{5}\frac{a^6}{(b^2-a^2)^3} - \ldots\right\},$$

so that when a is small compared with b,

$$\Delta = \frac{\rho m^2}{24\pi}\frac{a^2}{b^2}.$$

In this case the tension in the cylinder is very small compared with its value in the two previous cases: the value of Δ in the first case is greater than that in the second; the more the vortex filaments are concentrated at the axis the greater will be the value of Δ. Now let us suppose that a Faraday tube contains a given amount of vorticity distributed among irrotationally moving liquid; the axes of the vortex filaments being parallel to the axis of the tube. The moment of momentum of the fluid in the tube about its axis will depend upon the distribution of vorticity in the tube: the more the vorticity is concentrated near the axis of the tube the greater will be the moment of the momentum. Now suppose we apply a couple to the Faraday tube, the couple acting in such a direction as to increase the moment of momentum; this couple will cause the vortex filaments to concentrate more at the axis of the tube, and will consequently increase the tension in the tube. If, however, the couple on the Faraday tube acts in the opposite direction to the moment of momentum of the fluid in the tube, the action of the couple will cause the vortex filaments to spread out and get nearer the boundary of the tube: this will diminish the value of Δ, and consequently diminish the tension in the tube. If we suppose that the solid on which the tube abuts is an atom containing gyrostats, then when the gyrostats are rotating in the same direction as the fluid in the tube, we may regard the action between the gyrostats and the tube as equivalent to a couple tending to increase the moment of momentum of the fluid in the tube, and thus to increase the pull exerted by the tube on the atom. When, however, the rotation of the gyrostats is in the opposite direction to that of the fluid in the tube, the action between the atom and the tube will be equivalent to a couple tending to diminish the moment of momentum of the fluid in the tube, and thus to diminish the pull exerted by the tube on the atom. Thus if, as before, we suppose that the gyrostats in the hydrogen atom are rotating in the same direction as the fluid in the Faraday tube of which it is the origin when it carries a positive charge, whereas the gyrostats in the chlorine atom are rotating in the same direction as the fluid in the Faraday tube of which it is the termination when it carries a negative charge, we see that the attraction between a positively charged hydrogen atom and a negatively charged chlorine one will be greater than that between a negative hydrogen and a positive chlorine atom separated by the same distance.

The object of these illustrations is to call attention to the point that when charged atoms are close together, there may be forces partly electrical, partly chemical, in their origin in addition to those expressed by the ordinary laws of electrostatics.

There are one or two points in connexion with the theory of the electric field which can be illustrated by the conception of a Faraday tube as a bundle of vortex filaments, which, though not connected with the main object of this paper, may be briefly pointed out. The first of these arises from equation (4), p. 516, which indicates that the tension exerted by a vortex column is equal to the kinetic energy of the fluid in unit length of the column. Now we know that the forces on a charged body in the electric field are such as would be produced if there were a tension along the lines of force equal per unit area to the electrostatic energy in unit volume of the field. If we suppose the tension to be exerted by the Faraday tubes and the energy to reside in these tubes, this is equivalent to saying that the tension exerted by each of these tubes is equal to the energy in unit length of the tube. This exactly coincides with the result indicated by equation (4), if we suppose that the Faraday tubes are bundles of vortex filaments.

The other point is in connexion with the view that magnetic force is due to the movement of the Faraday tubes: the magnetic force being at right angles to the direction of the

Faraday tubes, and also to the direction in which they are moving, the magnitude of the force varying as the product of the "polarization" and the velocity of the tubes at right angles to their direction (see 'Recent Researches in Electricity and Magnetism,' by J. J. Thomson, p. 8). On this view the energy per unit volume in the magnetic field when the tubes are moving at right angles to themselves is (see 'Recent Researches,' p. 9)

$$\frac{\mu}{8\pi} P^2 V^2,$$

where μ is the magnetic permeability, P the polarization, and V the translatory velocity of the tubes. This expression would represent the kinetic energy due to the translatory motion of the tubes if the expression for the effective mass of the tubes contained a term proportional to the square of the "polarization." Now if we have a vortex column moving about in a fluid which is subject to other disturbances, the following considerations would seem to show that the expression for its effective mass would contain a term proportional to the square of the vorticity in the vortex column. The lines of flow when the vortex column is stationary in a

Fig. 2.

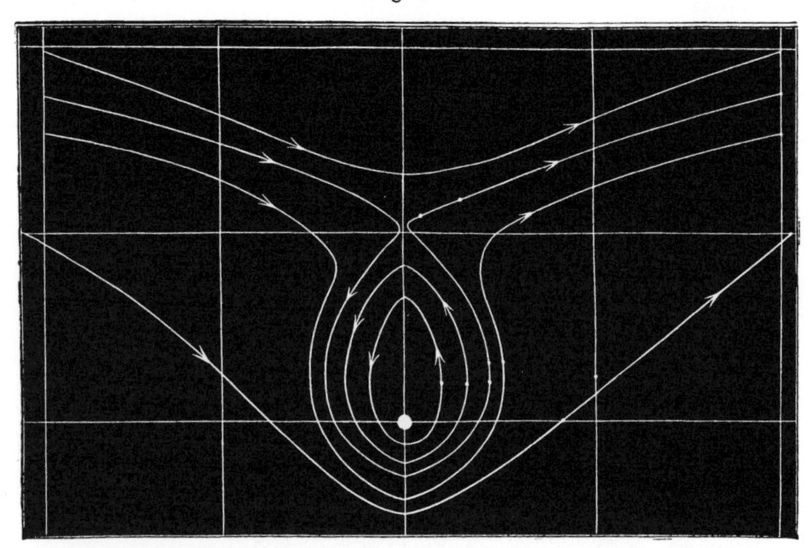

liquid moving so that at an infinite distance its velocity is uniform and horizontal are represented in fig. 2.

We see that some of these lines of flow in the neighbourhood of the column are closed curves ; now the liquid inside any one of these curves will always remain in the neighbourhood of the column, and if the column is moved will move with it : thus the effective mass of the column will be that of the column plus that of the liquid enclosed by the largest of the closed lines of flow. The linear dimensions of the curve are proportional to m/u, where u is the velocity of the fluid at an infinite distance from the column, and m the strength of the vortex ; (the equation to the bounding line of flow is easily seen to be $r=\frac{m}{u}e^{-\frac{uy}{m}}$), thus the area enclosed by the line of flow, and consequently the mass of fluid inside a cylinder of which it is the cross section, is proportional to m^2/u^2; thus, as the effective mass is increased by this mass of fluid, the expression for the effective mass of the vortex column will contain a term proportional to the square of the vorticity. Hence, if we regard a Faraday tube as a bundle of vortex filaments, we can by this analogy see that its effective inertia might involve a term proportional to the square of the polarization.

Relation of the preceding Analogies to the Electrochemical Properties of the Atoms.

To return, however, to the relation between the electric charge and the electrochemical properties of the element whose atom carries the charge. The illustration given on page 513 suggests that when an atom is charged with electricity it acquires a certain amount of potential energy depending upon the sign of the charge and also upon the kind of atom carrying the charge. Let us suppose that when an atom of an element A carries unit charge of positive electricity, its potential energy, in consequence of the connexion between the internal motion of the atom and the motion of the fluid in the Faraday tube, is greater by σ_A than when it has no charge, while when the atom has the unit negative charge its potential energy is less by σ_A than that of the uncharged atom. The quantity σ depends upon the nature of the atom ; in 'Recent Researches on Electricity and Magnetism,' p. 64, it is called the Volta potential of the substance, since the difference of potential between two metals A and B when placed in contact can be proved to be equal to $\sigma_A-\sigma_B$.

If the substance A has a charge Q of positive electricity, then in the expression for its potential energy there will be the term σ_AQ. If we consider this term alone, then if σ_A is positive an increase in the positive charge will involve an increase in the potential energy, while an increase in the negative charge would diminish the potential energy ; if σ_A is negative the converse is true. Now a dynamical system behaves so as to facilitate any change which causes a diminution in the potential energy. Thus those substances for which σ is negative will tend to acquire a charge of positive electricity, while those for which σ is positive will tend to get a charge of negative electricity. Thus the existence of the property expressed by the coefficient σ would produce the same effect as von Helmholtz's specific attraction of the elements for the two electricities.

Suppose, for example, that we have two metals A and B in contact, and that σ_A, σ_B are the values of the Volta coefficients for A and B respectively; let σ_A be greater than σ_B. Then if A acquires a negative charge equal to $-Q$, and B a positive charge equal to Q, the potential energy of the two metals will be diminished by $(\sigma_A - \sigma_B)Q$. If this were the only source of potential energy, the transference of positive electricity from A to B, and of negative from B to A, would go on indefinitely, as each transference would involve a diminution in the potential energy. The separation of the electricities will, however, produce an electric field the potential energy due to which will increase as Q increases, so that the diminution in energy due to the Volta effect will be accompanied by an increase in the energy due to the electrostatic field. Let us suppose, for example, that the two metals form the plates of a condenser whose capacity is C, then when A has a charge $-Q$ and B a charge $+Q$, the energy in the electrostatic field is equal to

$$\frac{1}{2}\frac{Q^2}{C},$$

while the energy due to the Volta effect is

$$-(\sigma_A - \sigma_B)Q.$$

The flow of negative electricity into A and of positive into B will go on until the increment in the energy of the electrostatic field due to an increase δQ in the charge on B is equal to the decrement of the energy due to the Volta effect produced by the same displacement of electricity.

Since the total energy is

$$\frac{1}{2}\frac{Q^2}{C} - (\sigma_A - \sigma_B)Q,$$

the increment when Q is increased by δQ is equal to

$$\delta Q\left\{\frac{Q}{C}-(\sigma_A-\sigma_B)\right\};$$

or, if V is the difference of potential between A and B, the increment of the potential energy is equal to

$$\delta Q\{V-(\sigma_A-\sigma_B)\}.$$

Thus so long as V is less than $\sigma_A-\sigma_B$, an increase in Q will be accompanied by a decrease in the potential energy, so that Q will tend to increase; while, on the other hand, when V is greater than $\sigma_A-\sigma_B$, an increase in Q will be accompanied by an increase in the potential energy, so that Q will tend to diminish: there will be equilibrium when $V=\sigma_A-\sigma_B$.

We see from this that if an atom of A has a unit negative charge, an atom of B a unit positive one, these atoms will retain their respective charges even though connected by a conductor, unless the potential of B exceeds that of A by more than $\sigma_A-\sigma_B$. If, on the other hand, A had a positive charge, B a negative one, they would, if connected by a conductor, interchange their charges unless the potential of B exceeded that of A by more than $\sigma_A-\sigma_B$.

Thus, assuming that σ is positive for chlorine and negative for hydrogen, an atom of hydrogen could retain a positive charge, and an atom of chlorine a negative one, even though the two were immersed in a conductor, provided the potential of the hydrogen atom did not exceed that of the chlorine atom by more than a certain limit; whereas if the hydrogen were negatively electrified and the chlorine positively, they would, if immersed in a conductor, interchange their charges unless the potential of the hydrogen atom exceeded that of the chlorine by the same limit as before.

Chemical Combination.

Thus, if we have a number of hydrogen atoms and an equal number of chlorine atoms immersed in a conductor, and if initially half of both the hydrogen and chlorine atoms were positively and half negatively electrified, interchange of charges between the atoms would go on until all the hydrogen atoms were positively and all the chlorine atoms negatively electrified.

For this interchange of charges to go on, however, it would seem necessary that a negatively electrified hydrogen atom and a positively electrified chlorine one should be connected by a conducting circuit. From the mechanical illustrations previously given, it seems unlikely that the atoms would interchange their charges by coming into contact, the positive charge passing from the chlorine to the hydrogen atom and *vice versâ*. From these illustrations we should rather expect that if a negative hydrogen atom came very near to a positive chlorine one, the two, if alone in the field, would tend to repel rather than to attract each other, the Faraday tube connecting the atoms ceasing to be straight, and bulging out into the surrounding medium somewhat in the manner shown in fig. 3.

Fig. 3.

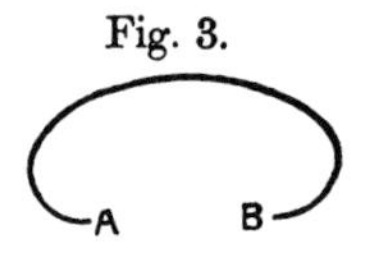

If in addition to the two atoms there were other charged bodies producing a very intense field tending to push the hydrogen and chlorine atoms together, then the interchange of their charges might take place by the ends of the Faraday tube gliding from one atom to the other after these atoms had been forced into contact by the external field. There are indications that the external field would have to be such as to produce a potential difference of a large number of volts between the two atoms before this method of exchanging their charges took place. In the absence of this potential difference, the atoms would not exchange their charges unless the medium into which the Faraday tube spread contained some conductor along which the ends of the Faraday tube could glide from one atom to the other. The necessity for this conducting circuit is perhaps one of the reasons why the presence of a third substance seems necessary for the continuance of many chemical reactions. How much a third substance able to act as a conductor could facilitate chemical combination may be seen from the following example. Suppose we have a mixture of hydrogen and chlorine molecules, and that by some external cause we split up these molecules into atoms; half of the hydrogen atoms produced by the dissociation of the hydrogen molecules will be positively electrified, while the other half will be negatively electrified; the same will also be true of the chlorine atoms. This condition will be permanent if the negative hydrogen atoms have no opportunities of interchanging their charges with the positive chlorine atoms, and a positive hydrogen atom would thus not be limited to combining with a negative chlorine one to form a molecule of hydrochloric acid, but might instead combine with a negative hydrogen atom to form a hydrogen molecule. If, however, conducting circuits, able to connect atom with atom, were present, the circumstances would be much more favourable to the formation of hydrochloric acid. For if the conducting circuit stretched from a positive hydrogen atom to a negative chlorine one, these atoms would retain their charges; whereas when the circuits stretched from a negatively electrified hydrogen atom to a positively electrified chlorine one, the atoms would interchange their charges. Thus the effect of these conducting circuits would be to cause all the hydrogen atoms to be positively electrified, and all the chlorine ones negatively ; this would of course increase the tendency for the hydrogen and chlorine to combine.

In the preceding case we have supposed the molecules to be already split up into atoms ; when, however, we consider the case of a mixture of molecules not already decomposed, we see that something more than the stretching of a conductor from one atom to another is required to effect the interchange of their charges. For suppose H, H, Cl, Cl (fig. 4) represent

Fig. 4.

```
H———H
+   −

Cl———Cl
+    −
```

respectively a hydrogen and a chlorine molecule, and suppose that the negative hydrogen atom is connected with the positive chlorine one by a conducting circuit. Then, if the negative charge of the hydrogen and the positive one of the chlorine were interchanged, the diminution in the potential energy due to the Volta effect would be $2(\sigma_{Cl}-\sigma_{H})$, where σ_{Cl}, σ_{H} are the Volta coefficients of chlorine and hydrogen respectively, σ_{H} being negative. To set off against this diminution in the potential energy due to the Volta effect, we have the increase in the energy produced by tearing the − charge on the H atom from its proximity to the + charge, and forcing it close to the − charge on the Cl atom. The increase in the potential energy due to this cause will be of the order $2(V_1+V_2)$, where V_1 and V_2 are the potential differences between the atoms in the hydrogen and chlorine molecules respectively. Thus the diminution in the potential energy when the charges are interchanged is

$$2(\sigma_{Cl}-\sigma_{H})-2(V_1-V_2),$$

and the interchange will not go on unless this quantity is positive. Now the potential difference due to the contact of two substances is equal to the difference of their Volta coefficients, so that the σ's will not exceed a small number of volts. To estimate the V's is more difficult, but we may remark that when we produce a spark through a gas, in which case there is strong evidence that we split up some of the molecules into atoms, then, no matter how short the spark may be, or what may be the pressure of the gas, the potential difference between the electrodes must exceed a certain value which is very large compared with the potential differences developed by the contact of heterogeneous substances, amounting in the case of hydrogen to between 190 and 200 volts. This minimum potential difference required to produce a spark is so constant under very varying physical conditions, such as pressure, spark-length, and so on, as to suggest that it represents some property of the molecule ; and I am inclined to think, and some experiments recently made at the Cavendish laboratory seem strongly to support the view, that the potential difference between the atoms in a molecule placed so as to be free from the action of other molecules is of the order of the minimum potential difference required to produce a spark. In the few cases where a direct estimate has been made of the work required to split up the molecule into atoms, such as that made by E. Wiedemann of the work required to decompose the hydrogen molecule, and that by Boltzmann for the iodine molecule, the potential difference indicated by these estimates far exceeds that produced by the contact of heterogeneous substances.

It would thus appear that in the case of gases where the molecules are free, the condition

$$2(\sigma_{Cl}-\sigma_{H}) > V_1 + V_2$$

is not fulfilled ; so that, on the electrical theory, chemical combination would not proceed. To produce chemical combination in such cases there must be some means of lowering the potential difference between the atoms in the molecules. Two methods by which this might take place at once suggest themselves. The first of these is that the combination, instead of taking place between a single pair of molecules, really takes place between aggregates of the molecules, physical aggregation preceding chemical combination.

Thus suppose that a number of molecules form themselves into a chain, such as that represented in fig. 5 ; then, if we consider a $-$H atom and the adjacent $+$Cl one, we see that the disposition of the charges on the atoms of the other molecules in the chain will diminish the work required to separate the $-$ charge on the H atom from the $+$ charge on the neighbouring H atom, and also that required to separate the + charge on the Cl atom from the − charge on the atom with which it is paired. Thus this disposition of molecules, by diminishing the electrostatic attractions, increases the chance for the −H and +Cl to interchange their charges in

Fig. 5.

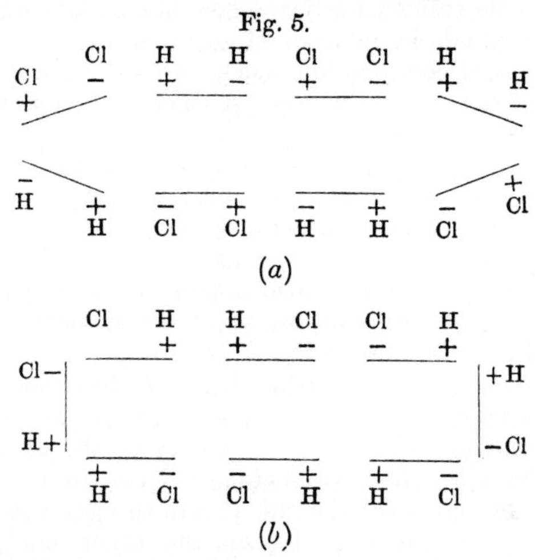

response to the "Volta effect:" after this interchange has taken place the adjacent H and Cl atoms will have respectively + and − charges, *i. e.* charges suitable for the formation of the HCl molecule; if such molecules form, the ring will be disposed as in fig. 5, *b*. This disposition will not be stable, since the lines joining the negative to the positive atoms in the molecule do not all point in one way; the ring will consequently break up into separate HCl molecules.

Thus, on this view the chemical combination consists in the formation of an aggregate of a large number of H and Cl molecules, then the interchange of the charges on some of the atoms, then the formation of an aggregate of HCl molecules, and finally the breaking up of this aggregate into a number of separate molecules of HCl. It will be noticed that there are no atoms set "free" during the whole of the process. The presence of free charged atoms during the progress of chemical combination between gases could be detected by experiments made on the electric conductivity of the mixture whilst the combination is proceeding. If free charged atoms are present they will move under electromotive forces, and will cause the mixture of gases to act as a conductor of electricity. A method of testing the conductivity of gases and the results of experiments made with it are given in a paper by the author (Proc. Camb. Phil. Soc. vol. viii. p. 258). These experiments showed that when a mixture of H and Cl was combining it did not possess any conductivity, thus proving that in these cases no charged atoms were set free during the combination. Some other cases of chemical combination gave the same result; on the other hand there were a class of cases in which the mixture of gases acted as a conductor whilst chemical combination was going on. The experiments thus proved that in many cases of chemical combination no charged atoms are set free.

I also investigated this point by another method. Let us assume for a moment that free charged atoms are liberated during chemical action. To fix our ideas, let us take the case of a piece of zinc attacked by chlorine; suppose now that we electrify the zinc: if chemical action liberated free ions, or, supposing that an aggregation of atoms is necessary, if the algebraical sum of the charges on the atoms forming this aggregate were not zero, then when the electrified zinc atom enters into chemical combination an atom of chlorine must be set free carrying a charge of electricity of the same sign as that on the electrified zinc atom. Thus, in this case the charge would leak from the metal into the gas and the latter would cease to insulate. If, on the other hand, chemical combination went on between an aggregate of atoms the algebraical sum of whose charges was zero, then, however much the zinc was attacked, there would be no liberation of charged atoms and no transference of electricity from the metal to the gas.

I made an extensive series of experiments to find out whether an electrified metal plate when immersed in a gas by which it was chemically attacked did or did not lose its charge. I was never able to trace any leakage of electricity to this cause, even though two metal plates charged to a great difference of potential, either by a Wimshurst machine or by a battery of 2000 small storage-cells, were placed close together in the gas. The electrified surfaces were zinc, mercury, and electrolytic solutions, the gases chlorine and nitrosyl chloride; but though the conditions of the surfaces showed that active chemical action had taken place, there was no escape of electricity through the gas whether the surfaces were positively or negatively electrified. In these cases the chemical combination went on at the temperature of the room when there was no chance of the gas being dissociated. In the case of combination at very high temperatures, when the gas may be dissociated, we shall see later that there are cases where chemical combination does promote the discharge of electricity.

2 O 2

5

X-rays and Cathode Rays (1895–1900)

Research Students

Two events in 1895 led to a rapid acceleration in the intensity and significance of Thomson's work on gaseous discharges. The first was an influx of new research students into the Cavendish following a change in the University regulations which permitted graduates of other universities to be admitted to undertake research leading to a degree. This coincided with a change in the regulations governing the award of scholarships by the Commissioners of the 1851 Exhibition which obliged award-holders to spend two years away from their home institution, resulting in a great mobility of promising young researchers around the British Empire. The second event was the discovery of X-rays by Röntgen.

The new policies had an immediate effect. Many dedicated young men from home and overseas began their research in the Cavendish, and Thomson's group expanded as a consequence. Included in the first cohort of students were three destined to achieve recognition: E. Rutherford, J.S.E. Townsend and J.A. McClelland. Rutherford came from Canterbury College, New Zealand, and pursued his research on wireless telegraphy using a detector he had invented before leaving that country. Soon after his arrival he succeeded in sending messages from the laboratory to his rooms, a distance of three-quarters of a mile – a record at the time.

Thomson had been a member of the University Committee which changed the regulations and played an important role in helping the new research students to feel at home in what must have seemed a strange, and at times unwelcoming, atmosphere. Many in Cambridge and the Cavendish were hostile to the newcomers, for they represented an elevation of the status of research within the University, and they competed for scarce resources.

Thomson, however, let all Cambridge know how he welcomed the newcomers and kept a watch over their interests. The daily ritual of tea was an opportunity for them to engage in social conversation and Thomson's presence always enlivened the atmosphere. Many students were later to recall their gratitude for his interest in their welfare as well as in their research.

Severe overcrowding in the laboratory soon followed and Thomson was faced with the problem of finding more room, both for research students and for the practical classes for medical students which had been instituted in 1888 and were temporarily housed in a tin hut in the back yard. These classes were vital to the Cavendish's economy: research students were supported and encouraged by being appointed to part-time demonstratorships, and by 1896 Thomson had accumulated £2000 from fees towards building an extension. The university contributed a further £2000. The extension was launched with a lavish reception for 700–800 guests, as Rutherford described to his fiancee Mary Newton:

> I told you it was to be a very big affair and no pains or expense was spared to make the thing a success. From the entrance to the Free School Lane to the Lab., about 60 yards, was covered in with an awning and lighted with glow lamps, and carpet laid down. Inside the Lab. itself carpet was laid down all over the staircases and everything was prettily lighted up . . . Mrs J J looked very well and was dressed very swagger and made a very fine hostess. J J himself wandered round looking very happy and grinning at everybody and everything in his own inimitable way. [1]

A further extension to the laboratory in 1908 was financed by a donation of £7000 from Lord Rayleigh's Nobel Prize money, and a further £2000 saved from fees.

The financial stringency imposed by all this saving was extreme. Robin Strutt recalled that

> The smallest expenditure had to be argued with him, and he was fertile in suggesting expedients by which it could be avoided – expedients which were more economical of money than of students time . . . Naturally, this financial stringency and the rapidly increasing number of workers in the laboratory created a severe competition for such apparatus as there was. A few who could afford to do so provided things of their own. Naturally the scarcity led to the development of predatory habits, and it was said that when one was assembling the apparatus for a research, it was necessary to carry a drawn sword in his right hand and the apparatus in his left. Someone moved an amendment – someone else's apparatus in his left. [2]

This statement is corroborated by Sir Lawrence Bragg's admission that,

> There was only one foot bellows between the forty of us for our glass blowing which we had to carry out for ourselves, and it was very hard to get hold of it. I managed to sneak it once from the room of a young lady researcher when she was

temporarily absent, and passing her room somewhat later I saw her bowed over her desk in floods of tears. I did not give the foot pump back. [3]

Yet the overcrowding brought benefits also and, led by Thomson's example, encouraged a notable cameraderie between researchers working on different, but related topics.

> . . . those who worked on the ground floor had frequent opportunities of talking to J J at various times of the day. He had no idea of reserving a time during which he was not to be interrupted. There are of course frequent intervals in experimental work when matters cannot be hurried – a suspected air leak in vacuum apparatus has to be given time to declare itself, or a glass apparatus recently made has to be given time to cool, and so on. The rooms on the ground floor all opened into one another, and their occupants wandered to and fro as they felt inclined. J J occupied the room at the end, and for a time his assistant, Everett, tried to establish the convention that it was private, indeed I think there was a notice to this effect on the door. But in practice little attention was paid to it, and when Rutherford, McLennan and others were established to work there as well as Everett, the game was up . . . The picture that remains . . . is of a score of individuals scattered about in various rooms, two or three in a larger room – but each working at his own particular problem . . . Glass work of very varying quality was usually conspicuous at bench level, with the ubiquitous Topler pump attached, and a maze of wires overhead: at least they should have been overhead, though I remember making a friendly protest to Townsend, who worked in the same room as I did, against his stretching wires in a position which threatened me with decapitation. [4]

An American visitor, Professor Bumstead of Yale, was struck by,

> . . . the (almost paradoxical) combination of qualities which I thought I observed [at the Cavendish]. It is obviously dominated by the personality of 'JJ' and yet I have never seen a laboratory in which there seemed to be so much independence and so little restraint on the man with ideas. The friendliness and mutual helpfulness of the research students was obvious and one of the finest things about the place, and it appeared to be a part of this friendly service to jump into a fellow-student, if you thought him wrong, and to prove him wrong. In a good many places friendship does not stand that strain, but it usually does at the Cavendish. [5]

Rose Thomson ably seconded J.J.'s interest in research student welfare, inviting them to teas and dinner, taking an interest in their personal lives, giving hospitality to their fiancees, and drafting them into helping with her pet charity, an annual entertainment for the people of Barnwell, a suburb of Cambridge.

In 1899 the Thomsons moved from Scrope Terrace to Holmleigh, on West Road, Cambridge, where the large garden was the main attraction to J.J. Initially the dining room doubled as his study and there was a tremendous upheaval every time he had to move out for their frequent dinner parties.

Their son George had been born in 1892, and later, in 1903, they had a daughter, Joan. An intervening child died at birth. Thomson used to go home to lunch, and this was the main time George saw him, perceiving him as 'a much loved but inscrutable Jove, mostly in the Olympian clouds of his own thoughts', [6] an account at odds with Rutherford's description of such a meal: 'Prof. J J is very fond of him [George] and played about with him during lunch while Mrs J J apologized for the informality.' [7]

X-rays

Professor W.C. Röntgen made his discovery of X-rays at the University of Wurzburg in November 1895, while investigating Lenard rays (cathode rays which appeared to have escaped from the discharge tube through a metallic window). To his surprise he observed fluorescence on a screen of paper covered with barium cyanide crystals two metres away from the apparatus. Röntgen suggested this was due to invisible 'X-rays' , which he showed were emitted from the glass wall of the tube where it was struck by cathode rays streaming off the negative electrode; subsequently he found that a target of platinum sealed inside the tube could also act as their source. The rays produced developable images on photographic plates and – the truly startling discovery – they penetrated skin and soft tissue to reveal, in such photographs, skeletal images of the hand and other parts of the human body. Within days of the publication of Röntgen's findings, X-ray mania gripped the world and raised public interest in the discovery to a degree that no other in physical science had done previously.

Thomson expounded on Röntgen's paper at a packed assembly of the Cavendish Physical Society. During the lecture, an X-ray photograph of a lady's hand was taken with one of the many replicas of Röntgen's tube which had been made by Everett.

> We organised a scheme at the Cavendish Laboratory by which photographs of patients brought by doctors were taken by my assistants, Mr Everett and Mr Hayles. The results were sometimes disconcerting to the doctors. . . . . One of the patients was a prominent member of the University who could express himself strongly. He had broken his arm and, as it did not heal, he insisted on having it Röntgen-rayed and brought his doctor with him. My assistant came back after a shorter time than usual. I asked him why. He said, 'I came away because I thought the doctor would not like me to hear what Mr ..... was saying to him after he saw the photograph. [8]

In the first months of 1896, Thomson published several papers on X-rays, reporting, among other things, two results of great significance for his work on discharge. The first of these was that X-rays discharged an electrified plate, irrespective of the sign of its charge. [9] The surrounding gas, or even an adjacent solid insulator, became conducting under exposure to the rays, allowing the charge to escape. In the case of gases, he found the rate of leak to be proportional to the square root of the pressure, as expected for a dissociation process mechanism in which each X-ray absorption event created two ions.

Physics Research Students June, 1898.

O.W.Richardson. J.Henry.

E.B.H.Wade. G.A.Shakespear. C.T.R.Wilson. E.Rutherford. W.Craig-Henderson. J.H.Vincent. G.B.Bryan.

J.C.McClelland. C.Child. P.Langevin. Prof.J.J.Thomson. J.Zeleny. R.S.Willows. H.A.Wilson. J.Townsend.

Research students at the Cavendish Laboratory

An early X-ray bulb made by Everett

Holmleigh. Thomson's home from 1899 until his appointment as Master of Trinity College

J.J., G.P. and Joan

Recalling his previous theories of ionisation by dissociation, Thomson stated 'The leakage of electricity through non-conductors is, I think, due to a kind of electrolysis, the molecule of the non-conductor being split up, or nearly split up, by the Röntgen rays, which act the part played by the solvent in ordinary electrolytic solutions.' [9]

Hitherto the only ways known of making electricity pass through a gas were either to apply large electric fields (of about 30 000 volts per centimetre) which frequently broke the discharge tubes or to use very hot gases such as occur in flames. Thomson found that when X-rays were directed onto a gas, a current flowed with the application of only a very small voltage. For the first time he had a readily controllable way of making a gas conduct. Moreover, the conducting gas was produced, and studied, outside the discharge tube from which the X-rays emanated. Thomson achieved, as he had attempted to do with electrodeless discharge, a separation between those effects which were due solely to the dissociation mechanism of conductivity, and those, such as the condition of the electrodes, which affected phenomena within the discharge tube. Now he could pinpoint and focus on the conduction mechanism and his theory of ionisation advanced rapidly.

The second discovery, reported in an important paper co-authored with McClelland, [10] was the observation of a saturation current in a gas made conducting by X-rays. At low applied voltages the current was found to be proportional to the voltage, but at higher values, the current approached a steady level. Thomson used this to underpin his dissociation theory. In a letter to Lord Kelvin (dated 10 April 1896) he wrote: 'The curve showing the relation between current and EMF is somewhat like that showing the relation between the magnetization of a piece of soft iron and the magnetic force. This opens up some interesting questions as to the nature of electrolysis generally.' [11]

It would appear that, by drawing on this analogy with magnetism – which Maxwell had suggested arose from the alignment of small particles in a magnetic field – Thomson had in mind his Grotthus-chain model of dissociation, the maximum current being reached when all these chains were aligned. Thomson also used the saturation current instrumentally, as a measure of the intensity of the X-rays, an important control in those days when the output of discharge tubes tended to be highly unstable.

Thomson soon exploited these two discoveries in a joint paper with Rutherford, [12] publication of which must stand as a milestone in his development of the theory of the electrical discharge. In it they achieved, for the first time, a mathematical formulation of the discharge by dissociation theory, which was in close quantitative agreement with experiment. It identified the important parameters of an ionised gas which were susceptible to quantitative measurement, and made definite, testable, predictions.

The experimental arrangement, although not illustrated in the paper, can be gleaned from the description given therein and from sketches in notebooks kept jointly by the two authors. One of these sketches is reproduced in Figure 5.1 along with an explanatory diagram.

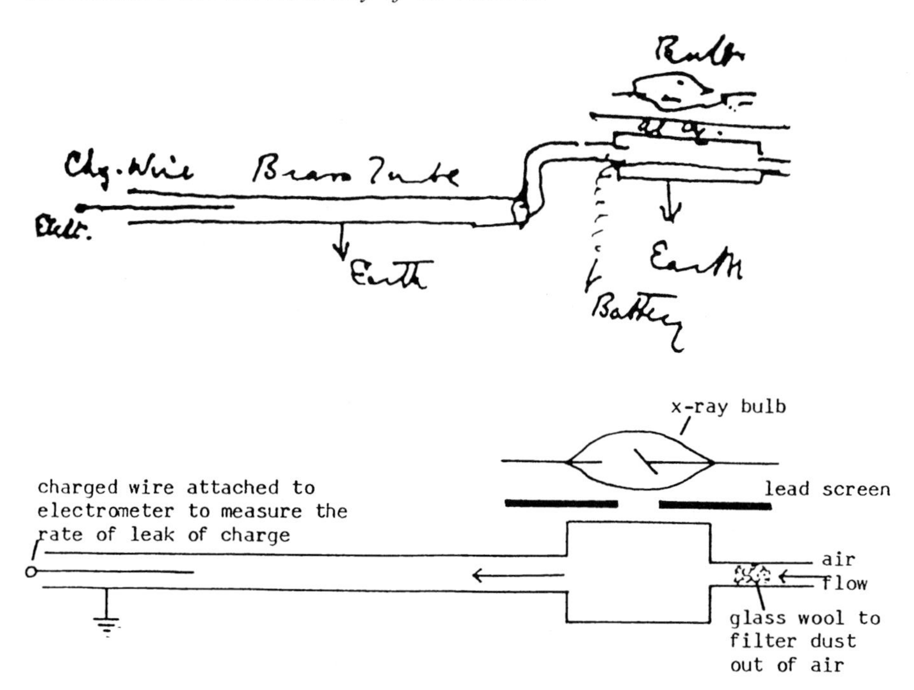

Figure 5.1 Sketch from Thomson and Rutherford's notebook illustrating the apparatus they used to study the ionisation of gases by X-rays (courtesy of Cambridge University Library [13]). The diagram below the sketch is from I. Falconer's PhD dissertation, University of Bath 1985

The essence of the first set of experiments was to expose a gas in the box on the right to X-rays and then to separate it from the source of ionisation by blowing it, with bellows, along the long tube and past a cylindical electrode arrangement. Any ionisation remaining in the gas after its passage down the tube could be sensed by the rate of leak of charge from the initially charged central electrode which was connected to a quadrant electrometer. They estimated that it took about half a second for the gas to reach the end of the tube, during which time the gas retained its conducting properties, i.e. its ionised state. They investigated whether the conductivity could be destroyed by passing the gas through a white-hot porcelain tube, by bubbling it through water, or by forcing it through a wire gauze, muslin or a porous plug of glass wool. These experiments imply that they were testing for the possibility of large aggregate, or chain-like ions. Heat did not destroy the conducting state, neither did passage through the gauze or muslin. However, the water and the glass wool were effective in so doing. Thomson and Rutherford concluded that 'the structure in virtue of which the gas conducts is of such a coarse character that it is not able to survive the passage through the fine pores in a plug of glass wool'. [12]

A significant result was obtained by passing an electric current through the gas, either in the X-ray ionisation chamber itself (as in the sketch, Figure 5.1) or at some point along the tube. Such a current, if of sufficient magnitude, was found to reduce or destroy the conducting state, as indicated by a marked reduction in the rate of leak from the wire electrode when the gas was blown past it. This discovery allowed a fundamental reinterpretation of the saturation current and opened ionisation up to quantitative investigation. Calling on Rutherford's previous experience of equilibrium equations, they now suggested that the saturation current represented an equilibrium between the creation of ions by the X-rays and their removal by the electric field. At low values of the applied voltage, the current, being carried by an essentially constant number of carriers, is proportional to the value of the electric field. However, at higher voltages the rate at which the carriers of electricity are removed at the electrodes by the field becomes comparable to the rate at which they are produced; the number of carriers then falls and the current tends towards saturation.

The equation governing the equilibrium could be expressed in terms of measurable experimental parameters such as the drift velocity of the ions and their recombination lifetime. Thomson and Rutherford went on to make numerous measurements of saturation current which they found depended on the gas used, with data being obtained for air, chlorine, hydrogen, coal gas, sulphuretted hydrogen and mercury vapour. An example of the plots for air and hydrogen is shown in Figure 5.2. From such plots they estimated values of the parameters (under the assumptions that both the charge and the velocity of negative and positive particles were equal), the linear portion of the I–V curves giving the product of the velocity and the lifetime, and the lifetime itself being measured in separate experiments in which the X-rays were cut off and the decay of current determined. They estimated the lifetime to be about one-tenth

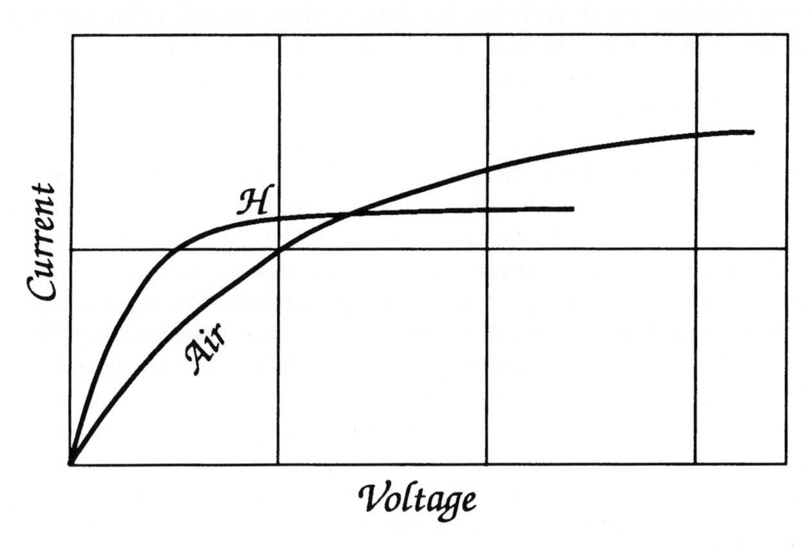

Figure 5.2 Current versus voltage characteristics associated with ionised air and hydrogen, as measured by Thomson and Rutherford

of a second and the velocity of the particles in air about 0.33 cm s^{-1} for an applied field of 1 volt cm^{-1}. They noted that the latter value is large compared with the velocity of ions in electrolysis but small compared with the velocity with which an atomic charge would move through a gas at atmospheric pressure (estimated to be about 50 cm s^{-1}). Much of the work of the Cavendish researchers over the next few years centred around measuring these parameters more reliably. Thomson and Rutherford concluded that the charged particles in an X-ray ionised gas 'are the centres of an aggregation of a considerable number of particles'.

The discovery of X-rays had stimulated renewed interest in cathode rays. In the autumn of 1896, with his confidence increased by his success with X-ray ionisation, Thomson turned to trying to assimilate cathode rays into his new theory.

The Cathode Ray Controversy

The observation of a fluorescent patch on the wall of a discharge tube which had been evacuated to a very low pressure was first made by Plucker. [14] The patch lay in a direction at right angles to the plane of the cathode, its position being independent of that of the anode, which could even be in a side arm of the tube. Furthermore, Plucker's student, Hittorf, showed that an object placed in the line of sight of the cathode cast a well defined shadow on the patch, suggesting the emission of rays which travelled in straight lines from the cathode. These cathode rays, as they were named, could be deflected by a magnetic field, thereby providing support for a description of their nature in terms of charged particles. However, it was not until 21 years later that Crookes, whose work on discharge phenomena created major interest at the time, made a forceful case for a particle description of cathode rays, specifically proposing that they were negatively charged molecules repelled at high velocity from the cathode. [15]

The particle view of cathode rays came into question when Hertz [16] found he could not deflect the rays with an electric field. This negative result added weight to the proposal, particularly advocated by Goldstein and by Wiedemann in 1880, that cathode rays were an etherial disturbance of some kind, similar to that of light. Although Goldstein's and Wiedemann's papers were later published in English in the *Philosophical Magazine*, [17] British scientists appeared to take little interest in what was to become a major controversy. However, when Hertz [18] found that cathode rays could pass through thin sheets of gold – an effect subsequently studied extensively by Lenard [19] after whom the transmitted rays were named – the advocates of the particle hypothesis faced a dilemma. It seemed inconceivable that particles the size of atoms or molecules could penetrate metal foils which were themselves composed of atoms. On the other hand, etherial waves could conceivably do so fairly easily.

Two other experimental results in the mid-1890s cast doubt on the etherial wave model and also influenced opinion. In 1894 Thomson modified his velocity of discharge experiment (see Chapter 4) to measure the velocity of cathode rays. His result – although later shown to be in error – gave a value substantially less than the velocity of light, and one year later Perrin [20] demonstrated that cathode rays entering a metallic cylinder through a small hole caused it to acquire a negative charge.

Rival schools now began to develop and, with the discovery of X-rays in 1896, the controversy concerning the nature of cathode rays took on an added dimension. Since X-rays were created by cathode rays, the intense interest generated by the former stimulated work on the latter. Some scientists attempted to provide a unified description of the two rays. For example, Batelli [21] put forward an ether-vortex theory of X-rays and suggested that it was probable that they were 'cathode rays sifted by the various media they have traversed'. Gifford [22] proposed that perhaps X-rays were polarised cathode rays and Vosmaer and Ortt [23] suggested they were particles which had lost their charge on passing through the walls of the discharge tube.

Nevertheless, the opinion of most was that X-rays were indeed etherial waves (although whether the vibrations were longitudinal or transverse had not been settled) and the controversy narrowed down to the problem of the nature of the cathode rays. Experimental investigations in 1897, resolved the problem in a conclusive manner.

Resolution of the Cathode Ray Problem

Thomson first announced his suggestion that cathode rays were *corpuscles*, small negatively charged particles from which atoms were built up, at a Friday evening discourse at the Royal Institution on 30 April 1897. It was published in *The Electrician*, [24] pre-dating his better known and fuller account in the *Philosophical Magazine* [25] by about six months.

The paper in *The Electrician* also appeared in *Proceedings of the Royal Institution* [24] which is the version reproduced in its entirety on p. 139. It may be said to represent Thomson's first account of the 'discovery of the electron', although it should be stressed that at this stage he chose not to use the term *electrons*, preferring his own term *corpuscles*. The reasons for this will be given in a subsequent section. Here we concentrate on the contents of this classic paper.

Following an historical introduction outlining previously reported properties of cathode rays, Thomson drew a contrast between the effect of a magnetic field on the trajectory of the spark discharge observed in gases at relatively high pressures with that found for the cathode rays which become prominent only in more highly evacuated tubes. The deflection of the former depends on the nature of the gas whereas the latter does not. The paths of the cathode rays were observed (and photographed) by the luminosity produced in the residual gas and by the phosphorescent patch on the glass where they struck the wall of the tube. Thomson found that the rays were deflected into an approximately circular path in just the manner expected for negatively charged particles. He was particularly struck by the fact that the deflection was the same whatever the gas in the discharge tube or the material of the electrodes. He confirmed that charges of negative electricity followed the course of the cathode rays using a modified form of Perrin's apparatus, but designed so as to counter the objection put forward by supporters of the etherial wave theory, namely that although 'electrified particles might be shot off from the cathode, these particles were, in their opinion, merely accidental accompaniments of the rays, and were no more to do with the rays than the bullet has with the flash of a rifle'. Thomson arranged that the collecting apparatus (a slit in a cylindrical anode leading to an inner cylinder connected to an electrometer) was not in the direct line of the rays, or of the supposed electrified particles. The apparatus is shown in Figure 10 of the paper. He then deflected the rays magnetically until they fell on the collecting slit. The inner cylinder acquired a large negative charge only when the rays passed through the slit, allowing Thomson to conclude that 'the stream of negatively electrified particles is an invariable accompaniment of the cathode rays'.

Other experiments reported in the paper were designed to elucidate the nature of Lenard rays: the apparent passage of cathode rays through a 1-mm-thick plate of brass within the discharge tube, and also their emergence, as Lenard had reported earlier, through a window to the outside of the tube. These phenomena Thomson attributed to the generation of an electric impulse by the arrival of the cathode rays on one side of the plate or window and the subsequent production of new cathode rays from the opposite side, rather than to penetration by the incident rays.

Thomson then made an important deduction concerning the size of the carriers of the negative electric charge based on the observations of Lenard that the cathode rays outside the tube travelled a distance of about half a centimetre before their intensity fell by one-half. From a comparison of this with the mean free path of molecules in air at atmospheric pressure (about 10^{-5} cm), Thomson concluded 'that the size of the carriers must be small compared with the dimensions of ordinary atoms or molecules'. He noted that the absorption was inversely proportional to the density of the gas and did not depend on the chemical nature of the gas. He continued,

> Let us trace the consequences of supposing that the atoms of the elements are aggregates of very small particles, all similar to each other; we shall call such particles corpuscles, so that the atoms of the ordinary elements are made up of corpuscles and holes, the holes being predominant. (see page 13 of the facsimile)

Thus he made the first announcement of the particle we now call the electron. He went on to describe how his theory would explain cathode rays:

> Let us suppose that at the cathode some of the molecules of the gas get split up into these corpuscles, and that these, charged with negative electricity, and mov-

> ing at a high velocity form the cathode rays . . . Now, the things these corpuscles strike against are other corpuscles, and not against the molecules as a whole; they are supposed to be able to thread their way between the interstices of the molecule . . . Thus the mean free path, and their coefficient of absorption, would depend only on the density (of the substance through which they pass); this is precisely Lenard's result. We see, too, on this hypothesis, why the magnetic deflection is the same inside the tube whatever be the nature of the gas, for the carriers of the charge are the corpuscles, and these are the same whatever gas be used. (see page 13 of the facsimile)

Thomson's suggestion was made solely on the basis of the independence of the magnetic deflection of the cathode rays of the nature of the gas or the material of the electrodes, and on the outstandingly low absorption of Lenard rays and its dependence on the mass of the absorbing gas. Moreover, the identification of Lenard rays with cathode rays was only scantily justified. Thomson was speculating far beyond the limits of his experimental data in suggesting the existence of corpuscles, and it is worth recalling how his previous ideas brought him to this conclusion.

Ever since his vortex atom theory of 1882, discussed in Chapter 2, he had been hypothesising about atoms with a structure depending on the arrangement of discrete units (originally vortices). Even when he abandoned vortex atoms around 1890, he retained an interest in structured atoms, suggesting in 1895 that the atom behaved as though it contained a number of outward-pointing gyroscopes (see Chapter 4).

However, the corpuscle idea required more than a structured atom. It required that the component particles were extremely small, and could exist independently of the atom. In 1895 Thomson had not arrived at either of these ideas, but his work on X-rays changed this situation.

On 10 April 1896, Thomson wrote to Kelvin describing his experiments with McClelland which seemed to show that the origin of X-rays was in the gas, rather than where the cathode rays hit the wall of the discharge tube. At this time he thought that X-rays were probably akin to light of very small wavelength. It was generally accepted that the visible spectra of elements were due to their molecules vibrating. Might the small wavelength of X-rays indicate that something much smaller than the molecule was vibrating?

> When there is more gas in the bulb the electricity passes through under a comparatively small potential difference, the gas gives out long waves (as shown by its luminosity) but no Röntgen waves [i.e. short waves]. When however the tube is giving out the R rays [sic] the potential difference required is very much greater than when the discharge is luminous. May not this great potential difference be able to produce a splitting up of the gaseous molecules into finer pieces, as it were, than before, the vibrations of these having wave lengths much smaller than those given out by the gas under the influence of the small potential difference in the ordinary Geissler tube? [26]

Although Thomson soon realised that the X-rays did actually originate from the glass wall of the discharge tube and abandoned this suggestion, the episode provided the germ of the idea of minute subatomic particles. Moreover, Röntgen had discovered that the absorption of X-rays by a gas was inversely proportional to the atomic weight of the gas, but independent of its chemical nature. By April 1896, Thomson had arrived at an explanation:

> This is a most interesting result and it may be remarked is what would occur on Prout's hypothesis of the constitution of the elements (that the elements were made up of aggregations of hydrogen atoms), if each little primordial atom furnished its quota to the absorption of these rays. [27]

Although in this case the constituents of the atom were atomic in size, this suggestion was very important to Thomson. A year later he realised that Lenard rays followed the same absorption rule, thus recalling the possibility of divisible atoms to his mind just as he was seeking an explanation for cathode rays. This, combined with the observation that Lenard rays travelled far further than a hydrogen atom possibly could before being absorbed, was quite enough to suggest a subatomic-sized particle to one who had already shown his willingness to speculate in this direction.

Thomson concluded his Royal Institution lecture with his first estimate of the ratio of the mass of the corpuscles to the charge they carry. It was based on a measurement of the heat generated by the cathode rays when they struck a thermal junction of known thermal capacity and of the radius of curvature of their path in a known magnetic field. The value obtained was $m/e = 1.6 \times 10^{-7}$, $\mathrm{g\,emu^{-1}}$, which Thomson immediately compared with the same ratio for the hydrogen atom, namely 10^{-4}. He concluded 'Taken . . . in conjunction with Lenard's results . . . these numbers seem to favour the hypothesis that the carriers of the charges are smaller than the atoms of hydrogen' (see page 14 of facsimile).

It is difficult to know how the audience at the Royal Institution, and later the readers of the paper, reacted to Thomson's revelations. Verification of the particle nature of cathode rays was probably surprising to some but the concept was not new. On the other hand, the suggestion that the particles were about one thousand times smaller than the size of the then smallest particle known, i.e. the hydrogen atom, must have raised eyebrows, and the even more revolutionary suggestion that the elements were built up of the corpuscles is likely to have been greeted with incredulity! The immediate reaction of FitzGerald [28] was that, if different elements are formed of the same kind of corpuscles, then one ought to be able to transmute any substance into any other by passing it through the furnace of cathode rays – the dream of the alchemists!

Any doubts that might have lingered as to the particle nature of cathode rays must have been dispelled when Thomson's description of new experiments were published later in the year. He showed that Hertz's earlier failure to obtain a deflection of the rays by application of an electric field was a conse-

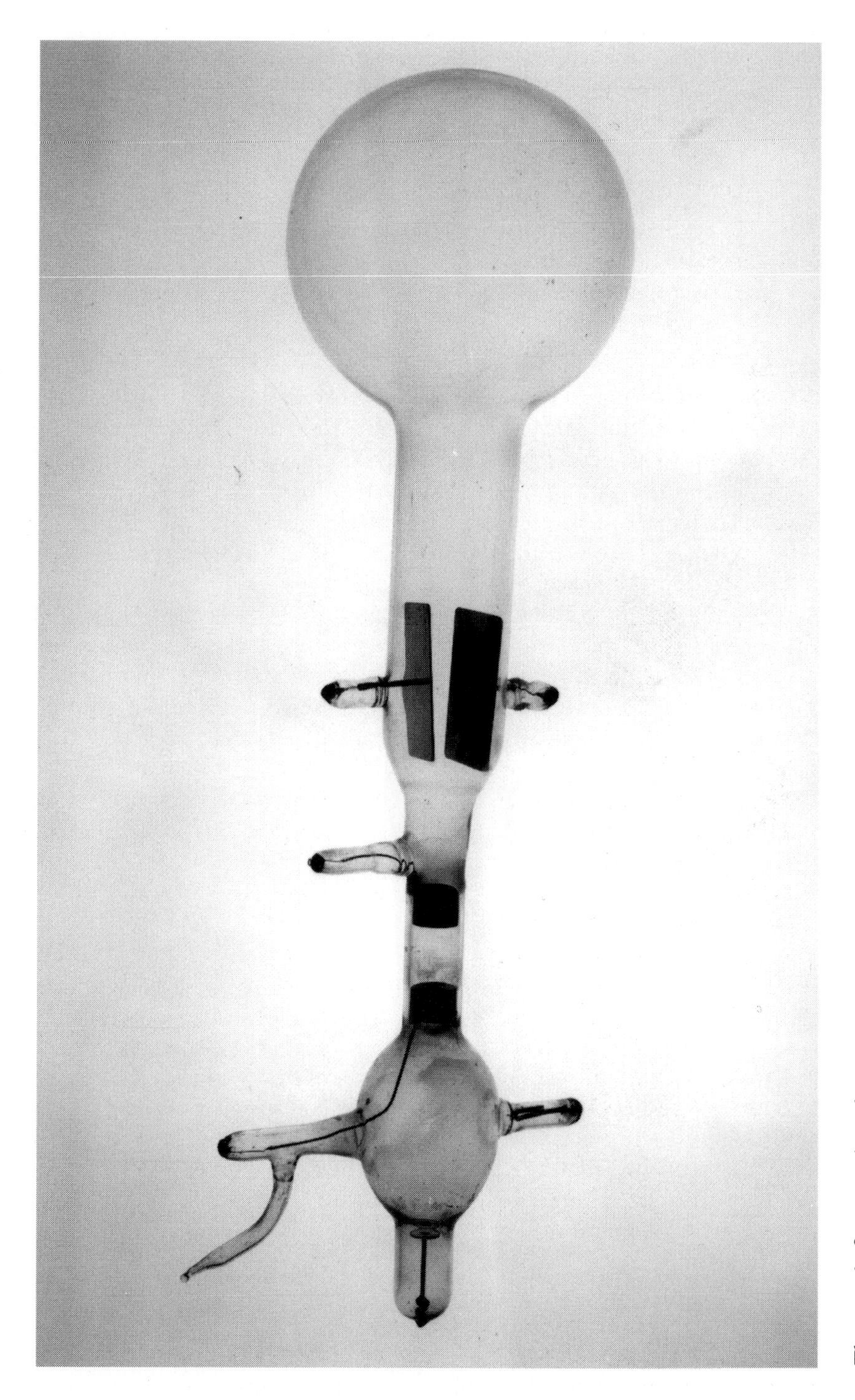

Thomson's famous e/m tube

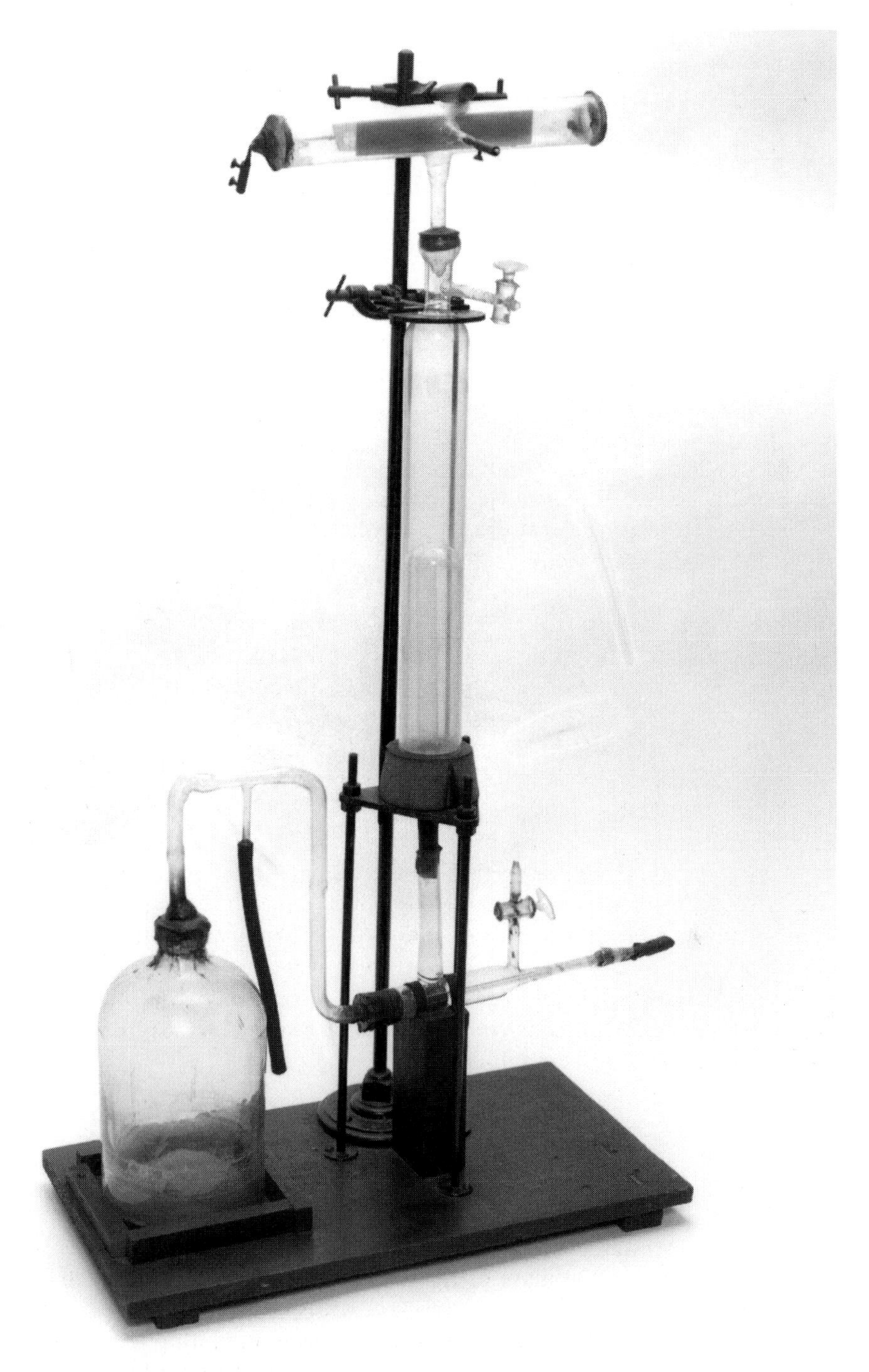

Expansion chamber used by Thomson to determine the charge of the corpuscles

quence of an insufficiently evacuated vessel. He had already, in 1893, suggested that the lack of electrostatic deflection was due to the conductivity of the residual gas in the discharge tube. By 1897, following his work on X-rays, he was well able to recognise the effects of ionisation and had observed a saturation in the amount of the charge collected in Perrin's experiment. Cathode rays evidently ionised the residual gas and caused it to conduct. He now spent several days evacuating his cathode ray tube, and finally obtained a deflection.

In an experiment that has since become the classic method of measuring the mass to charge ratio, Thomson used a new design of tube in which both electrostatic and magnetic deflections could be measured. The tube, whose design is similar to those used today in cathode ray oscilloscopes and television sets, is shown as Figure 2 of the paper in the *Philosophical Magazine* [29] a portion of which is reproduced on p. 153. A photograph of an original tube used by Thomson is shown on p. 127.

The new method of determining the m/e ratio did not necessitate measurement of the kinetic energy of the particles, which required assumptions about the transfer of energy from the rays to the thermocouple. It required only determinations of the electric and magnetic fields and the consequent deflection of the cathode rays.[1] Using different gases and metals for the electrodes (neither of which were expected to influence the results), Thomson reported values for m/e ranging between 1.1×10^{-7} and 1.5×10^{-7} $\mathrm{g\,emu}^{-1}$. As in his earlier paper, he contrasts this with the corresponding ratio for the hydrogen atom.[2]

In the current paper he is rather more circumspect about attributing the different ratios wholly to the small mass of the cathode ray corpuscles; indeed he states that 'the smallness of m/e may be due to the smallness of m or the largeness of e, or to a combination of these two.'[3]

Having demonstrated, more conclusively, that cathode rays were particles with a very low value of mass to charge ratio, Thomson reiterated his previous conclusion that these particles were universal constituents of atoms. He went on to propose an atomic structure based on corpuscles. He suggested that the corpuscles arranged themselves in rings in a repetitive way analogous to Mayer's magnets. The similarities between this and his 1882 vortex theory, discussed in Chapter 2, are remarkable.

Measurement of the Charge

Had Thomson chosen in 1897 to ascribe to the corpuscle a value of charge equal to that measured on certain ions in electrolysis experiments, he would have concluded that the mass of the corpuscle was indeed about 1000 times smaller than hydrogen. No doubt this idea did occur to him but it seems that he deliberately refrained from reaching this conclusion until after he had found a method of directly measuring the charge on the corpuscles themselves. This he was to do two years later.

The experimental technique used by Thomson to measure the charge carried by the corpuscles was first applied to a determination of the charge on ions produced in a gas by X-rays. [30] The principle was to measure the current through the ionised gas and to equate this to the product of the number of ions per unit volume, their charge and their mean velocity. For the velocity, Thomson used values obtained in earlier experiments by Rutherford on air and hydrogen (the two gases used), leaving the number of the ions as the only remaining quantity to be determined.

The method used to estimate the number of ions was both complicated and ingenious. The apparatus consisted of a vessel partly filled with water and containing the ionised gas. The vessel was connected to a vacuum-driven expansion device (shown on p. 128), operation of which resulted in condensation of water droplets around the ions that served as nuclei. It thus used the important discovery of Thomson's student, C.T.R. Wilson, that cooling produced by adiabatic expansion leads to condensation in ionised gases – a phenomenon which was later to be used to much effect in the Wilson cloud chamber. Thomson calculated the size of the drops from the rate at which they fell under gravity, using Stokes' equation for the terminal velocity of spherical objects falling in a viscous medium. Combining this with a measurement of the total amount of water deposited, he estimated the number of ions in a known volume.

The value Thomson finally obtained for the charge on the ions produced in air was 6.5×10^{-10} esu. Comparison with the 'usually given' charge carried by the hydrogen ion in electrolysis led him to conclude that the two were essentially the same. The numerical result (within 40 per cent of today's accepted value for the fundamental unit of charge, namely 4.8×10^{-10} esu) is extraordinary when it is considered how many different measurements and assumptions Thomson had to make in arriving at it. Determination of the total quantity of electricity flowing through the ionised gas necessitated use of a quadrant electrometer charged by Leclanché cells and was anything but straightforward. Measurement of the mass of water in the droplets involved knowledge of the degree of expansion and the temperature change. Other quantities needed were the latent heat of expansion of water, the specific heat, density and viscosity of the gas, and the velocity of the ions. Furthermore corrections were required for ions 'not caught by the expansion', droplets formed on dust and other nuclei, and the conductivity of the walls of the vessel.

One year after publication of the above results, Thomson reported an application of the same technique to the determination of the charge on the negatively charged particles produced when a metal is irradiated with ultraviolet light or a filament is heated to incandescence. He combined the method with a new technique for the determination of the m/e ratio of the particles and found that this had the same value as obtained earlier for the cathode ray corpuscles.

The historically significant paper [31] describing these results and conclusions is reproduced in full on p. 172. The experimental arrangement for the m/e experiment is illustrated in Figure 1 of the paper. A metal plate is positioned close to a wire grid in a vessel evacuated to a pressure of 1/100 mm of mercury. At this pressure the mean free path of air molecules is large and the electrified particles, emitted from the plate by ultraviolet radiation, travel under an applied potential gradient to the grid, the current being measured by a quadrant electrometer. Application of a magnetic field in the plane of the plate, in conjunction with the electric field perpendicular to it, causes the particles to execute a cycloid motion as illustrated in Figure 5.3. For high enough magnetic fields, the trajectory of the particles fails to intersect the grid, whereupon the current is expected to fall to zero. Thomson gives a simple theoretical treatment relating the diameter of the generating circle of the cycloids to the electric and magnetic fields – an expression that also contains the m/e ratio of the particles. Using various plate-to-grid separations, Thomson measured the magnetic fields at which the current fell sharply. From these he obtained an average value of e/m for the emitted particles of 7.3×10^6 emu g^{-1}. A somewhat higher value was found for particles emitted from an incandescent carbon filament. Nevertheless, Thomson considered both values were sufficiently close to each other and to the ratio he had obtained for the cathode ray particles to conclude that 'the particles which carry the negative electrification . . . are of the same nature' in all three cases. The importance of this conclusion is that the production of corpuscles had now been shown not to be solely confined to the gaseous discharge tube.

Thomson then proceeded to measure the charge on the particles produced by ultraviolet light. He used the method previously employed for ions produced by X-rays, i.e. a combination of measurements of the current carried by the particles and their density. The average of six measurements gave a value of 6.8×10^{-10} esu, which Thomson considered to be sufficiently close to the value of 6.5×10^{-10} (misprinted in the paper as 6.5×10^{-8}) deduced from the X-ray experiments to conclude that the charge was the same in both cases.

At the time of his cathode ray experiments, Thomson had for various reasons been reluctant to conclude that the small value of m/e he had found for the corpuscles necessarily implied a small value for m. In the later paper describing experiments on the charge of ions produced by X-rays, he remarked on the similarity of this charge with that on the hydrogen ion in electrolysis, but made no mention of any connection with the cathode ray corpuscles. In the present experiments, however, he clearly felt that measurements of both m/e and e for negative electrification produced by ultraviolet light, at last justified

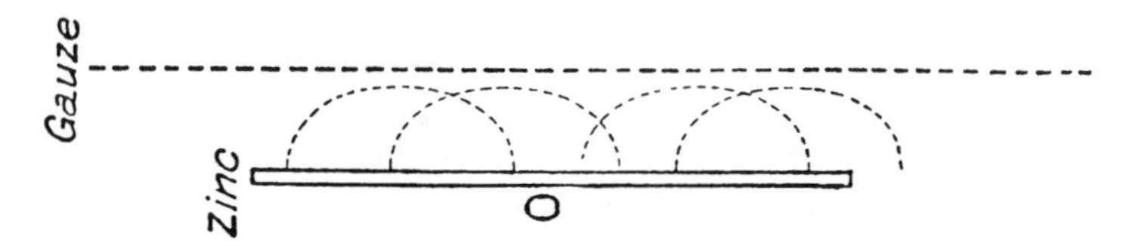

Figure 5.3 Illustration of cycloid motion of particles in an electric and a magnetic field

the deduction that the mass of the carriers was very small, namely about 1/1000 of that of the hydrogen atom. He writes:

> The experiments just described, taken in conjunction with previous ones on the value of m/e for the cathode rays, show that in gases at low pressures negative electrification, though it may be produced by very different means, is made up of units each having a charge of electricity of a definite size; the magnitude of this negative charge is about 6×10^{-10} electrostatic units, and is equal to the positive charge carried by the hydrogen atom in the electrolysis of solutions.
>
> In gases at low pressures these units of negative electric charge are always associated with carriers of a definite mass. This mass is exceedingly small, but only about 1.4×10^{-3} of that of the hydrogen ion, the smallest hitherto recognised as capable of a separate existence. The production of negative electrification thus involves the splitting up of an atom, as from a collection of atoms something is detached whose mass is less than that of a single atom.[4] (see page 563 of the facsimile)

Thomson's result was highly significant for it equated the charge of the corpuscle with those of Larmor's and Lorentz's electrons and thus linked it with an established body of electrodynamic theory. This opened up the possibility that all mass might be electromagnetic in origin. British physicists considered the experiment fundamental to electron physics, Oliver Lodge writing: 'it seems to me one of the most brilliant things that has recently been done in experimental physics. Indeed I should not need much urging to cancel the "recently" from this sentence.' [32]

Later in the paper, Thomson himself raises this question of whether the mass of the corpuscle is entirely due to its charge and suggests an experiment which might throw light on this point. The proposed experiment was a comparison of the heat produced by the particles when they strike the inside of a vessel composed of a substance transparent to X-rays with that produced when the vessel is opaque to these rays. He suggested that if the mass is 'mechanical', and not electrical, the heat produced should be the same in the two cases, whereas if the mass is electrical, the heat would be less in the first case because part of the energy would escape through the walls.

Towards the end of the paper, Thomson's confidence in the significance of his results is evident in bold statements concerning electrification of gases or indeed of matter in any state. Atoms are composed of a large number of smaller bodies (corpuscles) which he said have a mass of about 3×10^{-26} g.[5] Electrification involves detachment of a corpuscle from some of the atoms, the detached corpuscles behaving like negative ions and carrying a unit charge. The part of the atom left behind behaves as a positive ion with unit positive charge and a mass large compared with that of the negative ion. 'On this view, electrification essentially involves the splitting up of the atom, a part of the mass of the atom getting free and becoming detached from the original atom' (see page 565 of the facsimile).

It was this concept that the scientific community at large found difficult to accept.

Corpuscles versus Electrons

The term *electron* was introduced by Stoney [33] as the name for the fundamental unit of electric charge six years prior to Thomson's demonstration of the corpuscular nature of cathode rays. He coined it to describe the quantity of electricity traversing an electrolyte in electrolysis and the idea that electricity existed in separate natural units was advocated by others including Helmholtz. Concurrently, Larmor and Lorentz, independently developing Maxwell's electrodynamics, had been wrestling with the nature of electric charge. They had devised separate, but somewhat similar, theories which represented unitary charges, equal in value to those on the hydrogen ion, as strain centres in the ether. Such strain centres, which Larmor, following Stoney, began calling electrons in 1894, would have an associated electromagnetic mass. In 1896, Zeeman's discovery of the magnetic splitting of spectral lines gave the first indication of the likely sizes of the unitary charges. Zeeman calculated a mass to charge ratio for them of about 1/1000 that of the hydrogen ion. It might therefore be wondered why Thomson named the particle he had discovered a *corpuscle* rather than an *electron*, particularly after he had determined its charge and equated this with that on the hydrogen ion in electrolysis? Instead he deliberately avoided using the term electron for the particle until long after others began to adopt it.

The answer to this question resides in the meaning attached to the word electron. Following the theories of Larmor and of Lorentz, the electron was a particle of charge. But the corpuscle was more than this; it was an *essential* component of a divisible atom. It was a building block of matter as well as a particle carrying the same unit of negative electricity. Thomson evidently thought that the two concepts, electron and corpuscle, could and should remain essentially separate – even if related.

Others, however, were anxious to bring the two concepts together. FitzGerald's reaction to Thomson's first cathode ray paper (published in the same issue of *The Electrician*) has already been mentioned. FitzGerald certainly did not find Thomson's corpuscular theory of atoms convincing but accepted part of it. He proposed that the corpuscles, which he acknowledged Thomson had shown were small charged particles, were 'free electrons' – in the Larmor sense, i.e. charges, but incidental to the material atom. This proposal was of immense importance to Larmor and Lorentz. In cathode rays they had, for the first time, electrons in a form in which they could be isolated, manipulated and studied experimentally. Its significance to Lorentz is shown by his immediate recasting of his entire theory, which had hitherto been expressed in terms of collections of ions, into a form which treated individual electrons. He was thus able to determine experimentally the velocity dependence of the electron mass.

The ether theorists were also delighted with this proposal. They could accept that cathode rays were electrons, since these were perceived to be structures in the ether, possessing inertia but not true mass in the gravitational sense. It was therefore unnecessary for them to relinquish their previous position entirely.

Acceptance of Subatomic Particles

Thomson's suggestion that corpuscles were constituents of atoms was not generally acepted and he did not push for its acceptance until 1900. He left the corroboration of his theory largely to other people. Townsend [34] confirmed Thomson's measurement of the charge on the corpuscle and Kaufmann in Germany made numerous measurements of m/e and also of m itself. [35] Although Thomson was the sole recipient of the Nobel Prize for Physics in 1906 for his work on discharge in general, one nominee proposed that Thomson and Kaufmann share the award, and the Curies considered that Kaufmann's experiments were pre-eminent in confirming Thomson's theories about the mass of the corpuscle.

Two developments were of particular importance in the eventual acceptance of Thomson's theory. The first arose from Thomson's own suggestion that the mass of the corpuscle could be electromagnetic, just like that of the electron. Then, if as he supposed, atoms consisted entirely of corpuscles, matter itself could be regarded as an electromagnetic phenomenon. This idea appealed to theoretical physicists who were working towards an ultimate theory in which all matter was reduced to structures in the ether – an electromagnetic world picture as it were. The second development was a dawning realization from studies of radioactivity (discovered by Becquerel in 1896), that atoms could and did split up, even spontaneously, and change into different elements in the process. The charge of alchemy levelled at Thomson's model of the atom lost its force. Moreover, his model was the only one around that might explain how an atom might split up. An important link between the model and radioactivity was established when measurements of m/e for β rays showed that these were fast-moving corpuscles.

A final factor in what became a rapid and widespread acceptance of Thomson's corpuscular atom theory was the support he received from his students. These researchers, most of them young, represented the converted, who came to Cambridge specifically to collaborate with Thomson on his exciting ideas. John Zeleny, for example, was working in Berlin when Thomson announced his corpuscle hypothesis. Finding that no one else in Berlin believed in corpuscles, he packed his bags and came to Cambridge. Once there, Thomson inspired the researchers with his enthusiasm by 'his vital personality, his obvious conviction that what he and we were all doing was something important, and his camaraderie.' [36] They believed they were making history, for Thomson 'realised that what he had made was a revolution'. [36] Their confidence and mutual enjoyment found expression in the songs written for the Cavendish dinner, instituted in 1897. 'Ions Mine' was of composite authorship, Thomson himself contributing the fourth verse.

IONS MINE

Air: 'Clementine'

In the dusty lab'ratory'
'Mid the coils and wax and twine,
There the atoms in their glory
Ionise and recombine.

Chorus: Oh my darlings! Oh my darlings!
Oh my darling ions mine!
You are lost and gone for ever
When just once you recombine!

In a tube quite electrodeless,
They discharge around a line,
And the glow they leave behind them
Is quite corking for a time.

And with quite a small expansion,
1.8 or 1.9,
You can get a cloud delightful,
Which explains both snow and rain.

In the weird magnetic circuit
See how lovingly they twine,
As each ion describes a spiral
Round its own magnetic line.

Ultra-violet radiation
From the arc or glowing lime,
Soon discharges a conductor
If it's charged with minus sign.

α rays from radium bromide
Cause a zinc-blende screen to shine,
Set it glowing, clearly showing
Scintillations all the time.

Radium bromide emanation,
Rutherford did first divine
Turns to helium, then Sir William
Got the spectrum – every line.

The researchers' affection and admiration for Thomson, as well as his central position in the laboratory, is very evident in 'The Don of the Day' written by A.A. Robb:

THE DON OF THE DAY

Air: 'Father O'Flynn'

Of dons we can offer a charming variety,
All the big pots of the Royal Society,

Still there is no one of more notoriety
Than our professor, the pride of us all.
Here's a health to Professor J.J.
May he hunt ions for many a day,
  And take observations,
  And work out equations,
  And find the relations
  Which forces obey.

Chorus:
Here's a health to Professor J.J.
May he hunt ions for many a day,
  And take observations,
  And work out equations,
  And find the relations
  Which forces obey.

Our worthy professor is always devising
Some scheme that is startlingly new and surprising
In order to settle some question arising
On ions and why they behave as they do.
Thus when he wants to conclusively show
Some travel quickly and some travel slow,
  He brings into action
  Magnetic attraction
  And gets a deflection
  Above or below.

All preconceived notions he sets at defiance
By means of some neat and ingenious appliance,
By which he discovers a new law of science
Which no one had ever suspected before.
All the chemists went off into fits
Some of them thought they were losing their wits,
  When quite without warning
  (Their theories scorning)
  The atom one morning
  He broke into bits.

When the professor has solved a new riddle,
Or found a fresh fact, he's as fit as a fiddle,
He goes to the tea-room and sits in the middle
And jokes about everything under the sun.
Then if you try to look grave at his jest
You'll burst off the buttons which fasten your vest,
  For when he starts chaffing
  Though tea you be quaffing,
  You cannot help laughing
  Along with the rest.

When these researchers left Cambridge they took up academic posts all over the world, thus providing for a rapid dissemination of Thomson's work and support for it.

Notes

1 In modern classroom demonstrations of this classic experiment, the fields are normally applied simultaneously, their relative strengths being adjusted to give zero deflection of the rays. Contrary to many accounts, it does not appear from the paper that Thomson used this method.
2 Today's accepted value for m/e in the same units is 0.57×10^{-7} and so Thomson's estimate was a factor of about two too high.
3 The possibility of a large charge for the corpuscle was suggested to him by the additive properties of the specific inductive capacities of gases. It seems likely that his electrostatic deflection of cathode rays originated in a consequent attempt to measure their charge.
4 In later accounts of this work, an important point is often overlooked. In the experiment to determine the charge, the particles were *not* isolated corpuscles, as they were in the first experiment to measure the charge-to-mass ratio. The charge-determination experiment was conducted at a pressure close to atmospheric and the particles were *ions of atomic* size. This is clear from Thomson's use of a velocity (taken from Rutherford's work) appropriate to heavy ions. There is evidence in the paper that Thomson realised this distinction but it is not explicitly stated and, in places, the fact that the experiments were conducted on different entities is rather blurred. Usage by Thomson of the word *ion* to signify both species does not help to clarify the situation. Nevertheless, as long as the atomic-size ions carried only a single charge (which it is evident they did), then combination of the results of the two experiments to yield the mass of the isolated corpuscle was still justified.
5 Here Thomson may have made a mistake or the figure was misprinted. Taking his values for e/m of 7.3×10^6 emu g^{-1} and $e = 6.6 \times 10^{-10}$ esu gives $m \sim 3 \times 10^{-27}$ g, i.e. ten times smaller than quoted. Today's value for the mass of the electron is 9.11×10^{-28} g.

References

Cambridge University Library holds an important collection of Thomson manuscripts, classmark ADD 7654, referred to here as CUL ADD 7654, followed by the particular manuscript number. Other archives and a more complete bibliography, may be found in: Falconer, I., Theory and Experiment in J.J. Thomson's work on Gaseous Discharge, PhD Thesis, 1985, University of Bath.

[1] RUTHERFORD, E., Letter to Mary Newton, 16 March 1895, quoted in G. P. Thomson, *J. J. Thomson and the Cavendish Laboratory in his Day*, 1964, London: Nelson, pp. 87–88.

[2] Rayleigh, fourth Lord, *The Life of Sir J.J. Thomson*, 1942, Cambridge University Press. Reprinted 1969, London: Dawsons, p. 48.
[3] Quoted in Phillips, D., William Lawrence Bragg. *Biographical Memoirs of Fellows of the Royal Society*, **25** (1979) 8.
[4] See reference [2], pp. 49–51.
[5] *A History of the Cavendish Laboratory, 1871–1910*, 1910, London: Longmans, p. 225.
[6] Thomson, G. P., Manuscript autobiography, Trinity College Library, G P Thomson papers, A1 f16.
[7] See reference [1], pp. 90–91.
[8] Thomson, J. J., *Recollections and Reflections*, 1936, London: Bell. Reprinted 1975, New York: Arno, p. 402–3.
[9] Thomson, J. J., *Nature*, **53** (1896) 391.
[10] Thomson, J. J. and McClelland, J., *Proceedings of the Cambridge Philosophical Society*, **9** (1896) 126.
[11] CUL ADD 7342 T537.
[12] Thomson, J. J. and Rutherford, E., *Philosophical Magazine*, **42** (1896) 392.
[13] CUL ADD 7564 NB39.
[14] Plucker, *Annalen der Physik*, **103** (1858) 88.
[15] Crookes, W., *Philosophical Transactions of the Royal Society*, **170** (1879) 135, 641.
[16] Hertz, H., *Annalen der Physik*, **19** (1883) 782.
[17] Goldstein, E., *Philosophical Magazine*, **10** (1880) 173, 234; **14** (1882) 366. Wiedemann, E., *Philosophical Magazine*, **18** (1884) 35, 85.
[18] Hertz, H., *Electrician*, **32** (1894) 722.
[19] Lenard, P., *British Association for the Advancement of Science*, (1896) 709.
[20] Perrin, J., *Comptes rendus de l'Academie des Sciences, Paris*, **121** (1895) 1130.
[21] Batelli, *Nature*, **54** (1896) 62.
[22] Gifford, W., *Nature*, **54** (1896) 25 June.
[23] Vossmaer and Ortt, *Nature*, **56** (1897) 316.
[24] Thomson, J. J, *The Electrician*, **39** (1897) 104; also published in *Proceedings of the Royal Institution*, April 30, 1897, 1–14.
[25] Thomson, J. J., *Philosophical Magazine*, **44** (1897) 293.
[26] CUL ADD 7342 T537.
[27] Thomson, J. J., *Nature*, **53** (1896) 581.
[28] FitzGerald, G., *The Electrician*, **39** (1897) 103.
[29] Thomson, J. J., *Philosophical Magazine*, **44** (1897) 293.
[30] Thomson, J. J., *Philosophical Magazine*, **46** (1898) 528.
[31] Thomson, J. J., *Philosophical Magazine*, **48** (1899) 547.
[32] Lodge, P., *Electrons*, 1907, London: Bell, p. 76.
[33] Stoney, G. J., *Philosophical Transactions of the Royal Society*, **4** (1891) 518.
[34] Townsend, J. S., *Philosophical Transactions of the Royal Society A*, **193** (1900).
[35] Kaufmann, W., *Annalen der Physik*, **61** (1897) 544; **62** (1897) 596; *Gottingen Nachrichten*, **143** (1901).
[36] Richardson, O., Obituary of Thomson. *Nature*, **146** (1940) 355.

Royal Institution of Great Britain.

WEEKLY EVENING MEETING,

Friday, April 30, 1897.

Sir Frederick Bramwell, Bart. D.C.L. LL.D. F.R.S. Honorary Secretary and Vice-President, in the Chair.

Professor J. J. Thomson, M.A. LL.D. Sc.D. F.R.S.

Cathode Rays.

The first observer to leave any record of what are now known as the Cathode Rays seems to have been Plücker, who in 1859 observed the now well known green phosphorescence on the glass in the neighbourhood of the negative electrode. Plücker was the first physicist to make experiments on the discharge through a tube, in a state anything approaching what we should now call a high vacuum: he owed the opportunity to do this to his fellow townsman Geissler, who first made such vacua attainable. Plücker, who had made a very minute study of the effect of a magnetic field on the ordinary discharge which stretches from one terminal to the other, distinguished the discharge which produced the green phosphorescence from the ordinary discharge, by the difference in its behaviour when in a magnetic field. Plücker ascribed these phosphorescent patches to currents of electricity which went from the cathode to the walls of the tube, and then for some reason or other retraced their steps.

The subject was next taken up by Plücker's pupil, Hittorf, who greatly extended our knowledge of the subject, and to whom we owe the observation that a solid body placed between a pointed cathode and the walls of the tube cast a well defined shadow. This observation was extended by Goldstein, who found that a well marked, though not very sharply defined shadow was cast by a small body placed near a cathode of considerable area; this was a very important observation, for it showed that the rays casting the shadow came in a definite direction from the cathode. If the cathode were replaced by a luminous disc of the same size, this disc would not cast a shadow of a small object placed near it, for though the object might intercept the rays which came out normally from the disc, yet enough light would be given out sideways from other parts of the disc to prevent the shadow being at all well marked. Goldstein seems to have been the first to advance the theory, which has attained a good deal of prevalence in Germany, that these cathode rays are transversal vibrations in the ether.

The physicist, however, who did more than any one else to direct attention to these rays was Mr. Crookes, whose experiments, by their beauty and importance, attracted the attention of all physicists to this

B

subject, and who not only greatly increased our knowledge of the properties of the rays, but by his application of them to radiant matter spectroscopy has rendered them most important agents in chemical research.

Recently a great renewal of interest in these rays has taken place, owing to the remarkable properties possessed by an offspring of theirs, for the cathode rays are the parents of the Röntgen rays.

I shall confine myself this evening to endeavouring to give an account of some of the more recent investigations which have been made on the cathode rays. In the first place, when these rays fall on a substance they produce changes physical or chemical in the nature of the substance. In some cases this change is marked by a change in the colour of the substance, as in the case of the chlorides of the alkaline metals. Goldstein found that these when exposed to the cathode rays changed colour, the change, according to E. Wiedemann and Ebert, being due to the formation of a subchloride. Elster and Geitel have recently shown that these substances become photo-electric, i.e. acquire the power of discharging negative electricity under the action of light, after exposure to the cathode rays. But though it is only in comparatively few cases that the changes produced by the cathode rays shows itself in such a conspicuous way as by a change of colour, there is a much more widely spread phenomenon which shows the permanence of the effect produced by the impact of these rays. This is the phenomenon called by its discoverer, Prof. E. Wiedemann, thermoluminescence. Prof. Wiedemann finds that if bodies are exposed to the cathode rays for some time, when the bombardment stops the substance resumes to all appearance its original condition; when, however, we heat the substance, we find that a change has taken place, for the substance now, when heated, becomes luminous at a comparatively low temperature, one far below that of incandescence; the substance retains this property for months after the exposure to the rays has ceased. The phenomenon of thermoluminescence is especially marked in bodies which are called by Van t'Hoff solid solutions; these are formed when two salts, one greatly in excess of the other, are simultaneously precipitated from a solution. Under these circumstances the connection between the salts seems of a more intimate character than that existing in a mechanical mixture. I have here a solid solution of $CaSo_4$ with trace of $MnSo_4$, and you will see that after exposure to the cathode rays it becomes luminous when heated. Another proof of the alteration produced by these rays is the fact, discovered by Crookes, that after glass has been exposed for a long time to the impact of these rays, the intensity of its phosphorescence is less than when the rays first began to fall upon it. This alteration lasts for a long time, certainly for months, and Mr. Crookes has shown that it is able to survive the heating up of the glass to allow of the remaking of the bulb. I will now leave the chemical effects produced by these rays, and pass on to consider their behaviour when in a magnetic field.

' First, let us consider for a moment the effect of magnetic force on the ordinary discharge between terminals at a pressure much higher than that at which the cathode rays begin to come off. I have

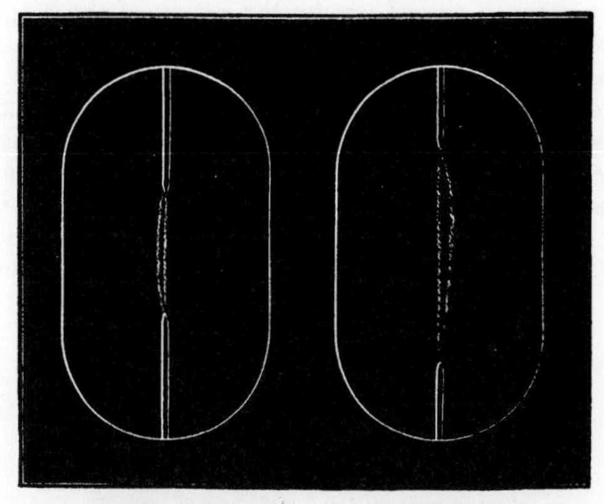

FIG. 1. FIG. 2.

here photographs (see Figs. 1 and 2) of the spark in a magnetic field. You see that when the discharge, which passes as a thin bright line between the terminals, is acted upon by the magnetic field, it is pulled aside as a stretched string would be if acted upon by a force at right

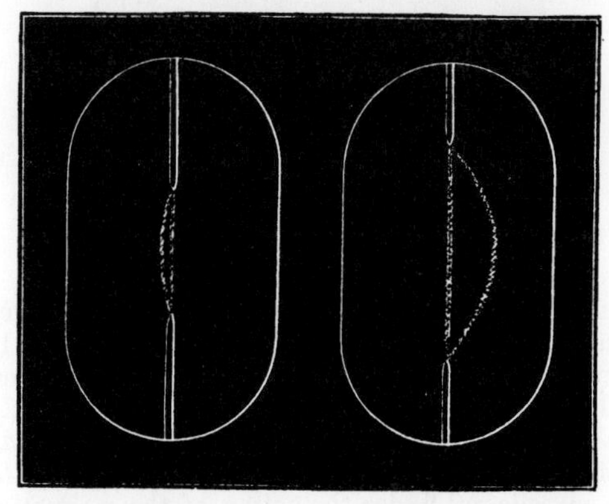

FIG. 3. FIG. 4.

angles to its length. The curve is quite continuous, and though there may be gaps in the luminosity of the discharge, yet there are no breaks at such points in the curve, into which the discharge is bent by a magnet. Again, if the discharge, instead of taking place between points, passes between flat discs, the effect of the magnetic force is to move the sparks as a whole, the sparks keeping straight until their terminations reach the edges of the discs. The fine thread-like discharge is not much spread out by the action of the magnetic field. The appearance of the discharge indicates that when the discharge passes through the gas it manufactures out of the gas something stretching from terminal to terminal, which, unlike a gas, is capable of sustaining a tension. The amount of deflection produced, other circumstances being the same, depends on the nature of the gas; as the photographs (Figs. 3 and 4) show, the deflection is very small in the case of hydrogen, and very considerable in the case of carbonic acid; as a general rule it seems smaller in elementary than in compound gases.

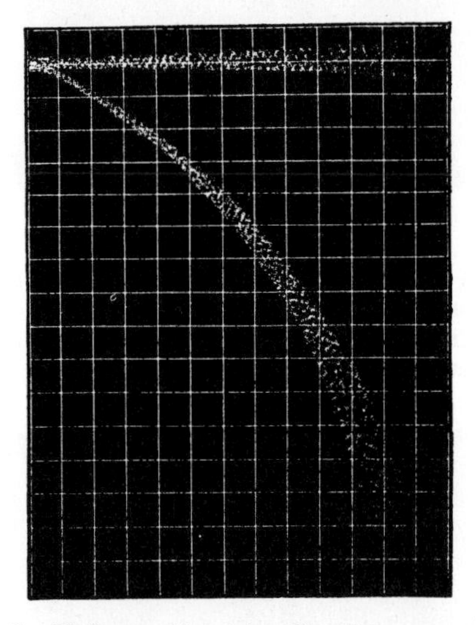

FIG. 5.—Hydrogen (Ammeter, 12; Voltmeter, 1600).

Let us contrast the behaviour of this kind of discharge under the action of a magnetic field with that of the cathode rays. I have here some photographs (Figs 5, 6 and 7) taken of a narrow beam formed by sending the cathode rays through a tube in which there was a plug with a slit in it, the plug being used as an anode and connected with the earth, these rays traversing a uniform magnetic field. The narrow beam spreads out under the action of the magnetic force into a broad fan-shaped luminosity in the gas. The luminosity in this fan is not uniformly distributed, but is condensed along certain lines. The phosphorescence produced when the rays reach the glass is also not uniformly distributed; it is much spread out, showing that the beam consists of rays which are not all deflected to the same extent by the magnet. The luminous patch on the glass is crossed by bands along which the luminosity is very much greater than in the adjacent

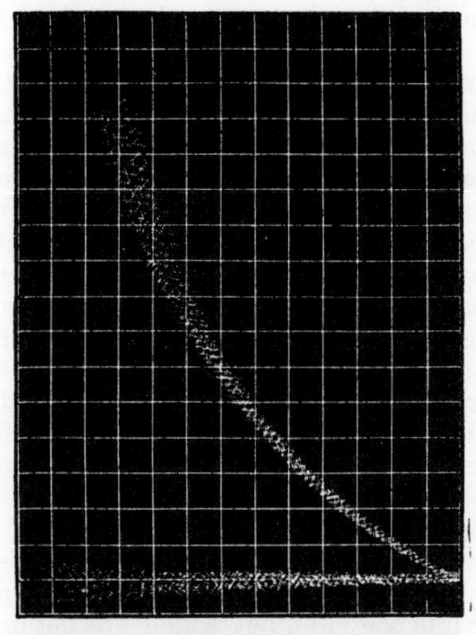

Fig. 6.—Air.

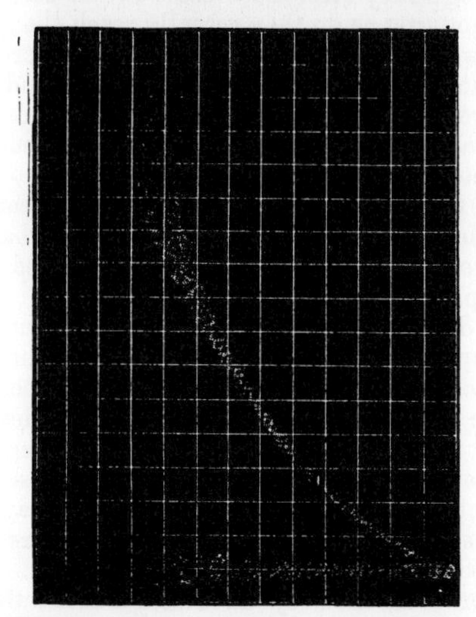

Fig. 7.—Carbonic Acid Gas (Ammeter, 12; Voltmeter, 1600).

parts. These bright and dark bands are called by Birkeland, who first boserved them, "the magnetic spectrum." The brightest places on the glass are by no means always the terminations of the brightest streaks of luminosity in the gas; in fact, in some cases a very bright spot on the glass is not connected with the cathode by any appreciable luminosity, though there is plenty of luminosity in other parts of the gas.

One very interesting point brought out by the photographs is that in a given magnetic field, with a given mean potential difference between the terminals, the path of the rays is independent of the nature of the gas; photographs were taken of the discharge in hydrogen, air, carbonic acid, methyl iodide, i.e. in gases whose densities range from 1 to 70, and yet not only were the paths of the most deflected rays the same in all cases, but even the details, such as the distribution of the bright and dark spaces, were the same; in fact, the photographs could hardly be distinguished from each other. It is to be noted that the pressures were not the same; the pressures were adjusted until the mean potential difference was the same. When the pressure of the gas is lowered, the potential difference between the terminals increases, and the deflection of the rays produced by a magnet diminishes, or at any rate the deflection of the rays where the phosphorescence is a maximum diminishes. If an air break is inserted in the circuit an effect of the same kind is produced. In all the photographs of the cathode rays one sees indications of rays which stretch far into the bulb, but which are not deflected at all by a magnet. Though they stretch for some two or three inches, yet in none of these photographs do they actually reach the glass. In some experiments, however, I placed inside the tube a screen, near to the slit through which the cathode rays came, and found that no appreciable phosphorescence was produced when the non-deflected rays struck the screen, while there was vivid phosphorescence at the places where the deflected rays struck the screen. These non-deflected rays do not seem to exhibit any of the characteristics of cathode rays, and it seems possible that they are merely jets of uncharged luminous gas shot out through the slit from the neighbourhood of the cathode by a kind of explosion when the discharge passes.

The curves described by the cathode rays in a uniform magnetic field are, very approximately at any rate, circular for a large part of their course; this is the path which would be described if the cathode rays marked the path of negatively electrified particles projected with great velocities from the neighbourhood of the negative electrode. Indeed, all the effects produced by a magnet on these rays, and some of these are complicated, as, for example, when the rays are curled up into spirals under the action of a magnetic force, are in exact agreement with the consequences of this view.

We can, moreover, show by direct experiment that a charge of negative electricity follows the course of the cathode rays. One way in which this has been done is by an experiment due to Perrin, the details of which are shown in the accompanying figure (Fig. 8.) In this experiment the rays are allowed to pass inside a metallic cylinder through a small hole, and the cylinder, when these rays enter it, gets a negative charge, while if the rays are deflected by a magnet, so as to escape the hole, the cylinder remains without charge. It seems to me that to the experiment in this form it might be objected that, though the experiment shows that negatively electrified bodies are projected normally from the cathode, and are deflected by a magnet, it does not show that when the cathode rays are deflected by a magnet the path of the electrified particles coincides with the path of the cathode rays. The supporters of the theory that these rays are waves

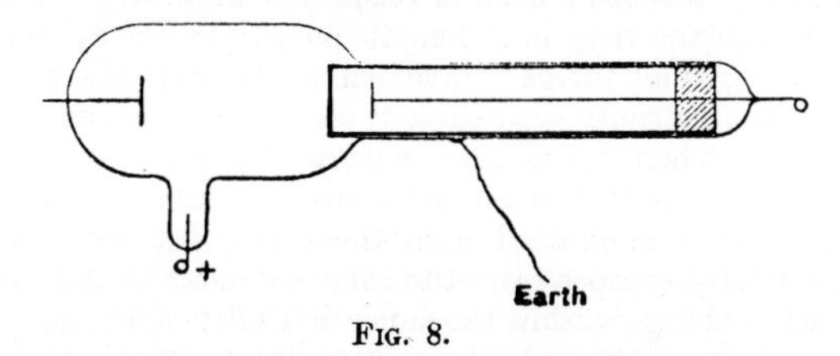

FIG. 8.

in the ether might say, and indeed have said, that while they did not deny that electrified particles might be shot off from the cathode, these particles were, in their opinion, merely accidental accompaniments of the rays, and were no more to do with the rays than the bullet has with the flash of a rifle. The following modification of Perrin's experiment is not, however, open to this objection: Two co-axial cylinders (Fig. 9), with slits cut in them, the outer cylinder being connected with earth, the inner with the electrometer, are placed in the discharge tube, but in such a position that the cathode

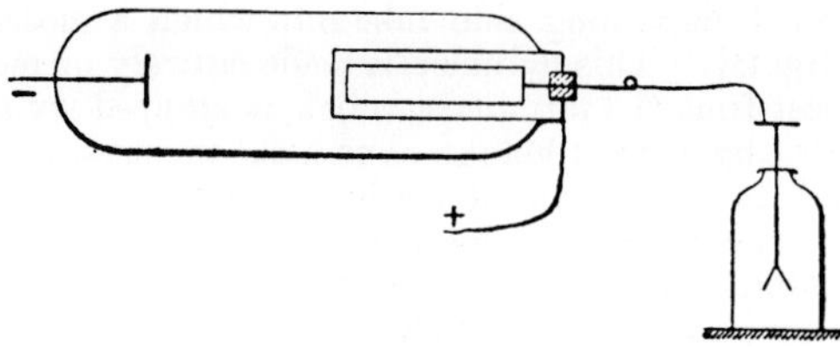

FIG. 9.

rays do not fall upon them unless deflected by a magnet; by means of a magnet, however, we can deflect the cathode rays until they fall on the slit in the cylinder. If under these circumstances the cylinder gets a negative charge when the cathode rays fall on the slit, and remains uncharged unless they do so, we may conclude, I think, the stream of negatively-electrified particles is an invariable accompaniment of the cathode rays. I will now try the experiment. You notice that when there is no magnetic force, though the rays do not fall on the cylinder, there is a slight deflection of the electrometer, showing that it has acquired a small negative charge. This is, I think, due to the plug getting negatively charged under the torrent of negatively electrified particles from the cathode, and getting out cathode rays on its own account which have not come through the slit. I will now deflect the rays by a magnet, and you will see that at first there is little or no change in the deflection of the electrometer, but that when the rays reach the cylinder there is at once a great increase in the deflection, showing that the rays are pouring a charge of negative electricity into the cylinder. The deflection of the electrometer reaches a certain value and then stops and remains constant, though the rays continue to pour into the cylinder. This is due to the fact that the gas traversed by the cathode rays becomes a conductor of electricity, and thus, though the inner cylinder is perfectly insulated when the rays are not passing, yet as soon as the rays pass through the bulb the air between the inner cylinder and the outer one, which is connected with the earth, becomes a conductor, and the electricity escapes from the inner cylinder to the earth. For this reason the charge within the inner cylinder does not go on continually increasing: the cylinder settles into a state of equilibrium in which the rate at which it gains negative electricity from the rays is equal to the rate at which it loses it by conduction through the air. If we charge up the cylinder positively it rapidly loses its positive charge and acquires a negative one, while if we charge it up negatively it will leak if its initial negative potential is greater than its equilibrium value.

I have lately made some experiments which are interesting from the bearing they have on the charges carried by the cathode rays, as well as on the production of cathode rays outside the tube. The experiments are of the following kind. In the tube (Fig. 10) A and B are terminals. C is a long side tube into which a closed metallic cylinder fits lightly. This cylinder is made entirely of metal except the end furthest from the terminals, which is stopped by an ebonite plug, perforated by a small hole so as to make the pressure inside the cylinder equal to that in the discharge tube. Inside the cylinder there is a metal disc supported by a metal rod which passes through the ebonite plug, and is connected with an electrometer, the wires making this connection being surrounded by tubes connected with the earth so as to screen off electrostatic induction. If the end of the cylinder is made of thin aluminium about $\frac{1}{20}$th of a millimeter thick, and a discharge sent between the terminals, A being the cathode, then at pressures far higher than those at which the cathode rays come off, the disc inside the cylinder acquires a positive charge. And if it is charged up independently the charge leaks away, and it leaks more rapidly when the disc is charged negatively than when it is charged positively; there is, however, a leak in both cases, showing that conduction has taken place through the gas between the cylinder and the disc. As the pressure in the tube is diminished the positive charge on the disc diminishes until it becomes unappreciable. The leak from the disc when it is charged still continues, and is now equally rapid, whether the original charge on the disc is positive or negative. When the pressure falls so low that cathode rays begin to fall on the end of the cylinder, then the disc acquires a negative charge, and the leak from the disc is more rapid when it is charged positively than when it is charged negatively. If the cathode rays are pulled off the end of the cylinder by a magnet, then the negative charge on the disc and the rate of leak from the disc when it is positively charged is very much diminished. A very interesting point is that these effects, due to the cathode rays, are observed behind comparatively thick walls. I have here a cylinder whose base is brass about 1 mm. thick, and yet when this is exposed to the cathode rays the disc behind it gets a negative charge, and leaks if charged positively. The effect is small compared with that in the cylinder with the thin aluminium base, but is quite appreciable. With the cylinder with the thick end I have never been able to observe any effect at the higher pressures when no cathode rays were coming off. The effect with the cylinder with the thin end was observed when the discharge was produced by a large number of small storage cells, as well as when it was produced by an induction coil.

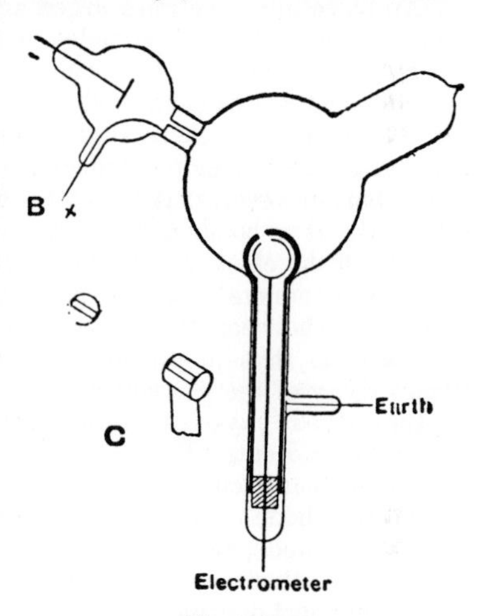

FIG. 10.

It would seem from this experiment that the incidence of the cathode rays on a brass plate as much as 1 mm. thick, and connected with the earth, can put a rarified gas shielded by the plate into a condition in which it can conduct electricity, and that a body placed behind this screen gets a negative charge, so that the side of the brass away from the cathode rays acts itself like a cathode though kept permanently to earth. In the case of the thick brass the effect seems much more likely to be due to a sudden change in the potential of the outer cylinder at the places where the rays strike, rather than to the penetration of any kinds of waves or rays. If the discharge in the tube was perfectly continuous the potential of the outer cylinder would be constant, and since it is connected to earth by a wire through which no considerable current flows, the potential must be approximately that of the earth. The discharge there cannot be continuous; the negative charge must come in gusts against the ends of the cylinder, coming so suddenly that the electricity has no time to distribute itself over the cylinder so as to shield off the inside from the electrostatic action of the cathode rays; this force penetrates the cylinder and produces a discharge of electricity from the far side of the brass.

Another effect which I believe is due to the negative electrification carried by the rays is the following. In a very highly exhausted tube provided with a metal plug, I have sometimes observed, after the coil has been turned off, bright patches on the glass; these are deflected by a magnet, and seem to be caused by the plug getting such a large negative charge that the negative electricity continues to stream from it after the coil is stopped.

An objection sometimes urged against the view that these cathode rays consist of charged particles, is that they are not deflected by an electrostatic force. If, for example, we make, as Hertz did, the rays pass between plates connected with a battery, so that an electrostatic force acts between these plates, the cathode ray is able to traverse this space without being deflected one way or the other. We must remember, however, that the cathode rays, when they pass through a gas make it a conductor, so that the gas acting like a conductor screens off the electric force from the charged particle, and when the plates are immersed in the gas, and a definite potential difference established between the plates, the conductivity of the gas close to the cathode rays is probably enormously greater than the average conductivity of the gas between the plates, and the potential gradient on the cathode rays is therefore very small compared with the average potential gradient. We can, however, produce electrostatic results if we put the conductors which are to deflect the rays in the dark space next the cathode. I have here a tube in which, inside the dark space next the cathode, two conductors are inserted; the cathode rays start from the cathode and have to pass between these conductors; if, now, I connect one of these conductors to earth there is a decided deflection of the cathode rays, while if I connect the other electrode to earth there is a deflection in the opposite direction. I ascribe this deflection to the gas in the dark space either not being a conductor at all, or if a conductor, a poor one compared to the gas in the main body of the tube.

Goldstein has shown that if a tube is furnished with two cathodes, when the rays from one cathode pass near the other they are repelled from it. This is just what would happen if the dark space round the electrode were an insulator, and so able to transmit electrostatic attractions or repulsions. To show that the gas in the dark space differs in its properties from the rest of the gas, I will try the following experiment. I have here two spherical bulbs connected together by a glass tube; one of these bulbs is small, the other large; they each contain a cathode, and the pressure of the gas is such that the dark space round the cathode in the small bulb completely fills the bulb, while that round the one in the larger bulb does not extend to the walls of the bulb. The two bulbs are wound with wire, which connects the outsides of two Leyden jars; the insides of these jars are connected with the terminals of a Wimshurst machine. When sparks pass between these terminals currents pass through the wire which induce currents in the bulbs, and cause a ring discharge to pass through them. Things are so arranged that the ring is faint in the larger bulb, bright in the smaller one. On making the wires in these bulbs cathodes, however, the discharge in the small bulb, which is filled by the dark space, is completely stopped, while that in the larger one becomes brighter. Thus the gas in the dark space is changed, and in the opposite way from that in the rest of the tube. It is remarkable that when the coil is stopped the ring discharge on both bulbs stops, and it is some time before it starts again.

The deflection excited on each other by two cathodic streams would seem to have a great deal to do with the beautiful phosphorescent figures which Goldstein obtained by using cathodes of different shapes. I have here two bulbs containing cathodes shaped like a cross; they are curved, and of the same radius as the bulb, so that if the rays came off these cathodes normally the phosphorescent picture ought to be a cross of the same size as the cathode, instead of being of the same size. You see that in one of these bulbs the image of the cross consists of two large sectors at right angles to each other, bounded by bright lines, and in the other, which is at a lower pressure, the geometrical image of the cross, instead of being bright, is dark, while the luminosity occupies the space between the arms of the cross.

So far I have only considered the behaviour of the cathode rays inside the bulb, but Lenard has been able to get these rays outside the tube. To this he let the rays fall on a window in the tube, made of thin aluminium about $\frac{1}{100}$th of a millimetre thick, and he found that from this window there proceeded in all directions rays which were deflected by a magnet, and which produced phosphorescence when they fell upon certain substances, notably upon tissue paper soaked in a solution of pentadekaparalolylketon. The very thin aluminium is difficult to get, and Mr. McClelland has found that if it is not necessary to maintain the vacuum for a long time, oiled silk answers admirably for a window. As the window is small the phosphorescent patch produced by it is not bright, so that I will show instead the other property of the cathode rays, that of carrying with them a negative charge. I will place this cylinder in front of the hole, connect it with the electrometer, turn on the rays, and you will see the cylinder gets a negative charge; indeed this charge is large enough to produce the well known negative figures when the rays fall on a piece of ebonite which is afterwards dusted with a mixture of red lead and sulphur.

From the experiments with the closed cylinder we have seen that when the negative rays come up to a surface even as thick as a millimetre, the opposite side of that surface acts like a cathode, and gives off the cathodic rays; and from this point of view we can understand the very interesting result of Lenard that the magnetic deflection of the rays outside the tube is independent of the density and chemical composition of the gas outside the tube, though it varies very much with the pressure of the gas inside the tube. The cathode rays could be started by an electric impulse which would depend entirely on what was going on inside the tube; since the impulse is the same the momentum acquired by the particles outside would be the same; and as the curvature of the path only depends on the momentum, the path of these particles outside the tube would only depend on the state of affairs inside the tube.

The investigation by Lenard on the absorption of these rays shows that there is more in his experiment than is covered by this consideration. Lenard measured the distance these rays would have to travel before the intensity of the rays fell to one-half their original value. The results are given in the following table:—

Substance.	Coefficient of Absorption.	Density.	$\frac{\text{Absorption}}{\text{Density}}$.
Hydrogen (3 mm. press.)	0·00149	0·000000368	4040
" (760)	0·476	0·0000484	5640
Air (0·760 mm. press.)	3·42	0·00123	2780
SO_2	8·51	0·00271	3110
Collodion	3,310	1·1	3010
Glass	7,810	2·47	3160
Aluminium	7,150	2·70	2650
Silver	32,200	10·5	3070
Gold	53,600	19·3	2880

We see that though the densities and the coefficient of absorption vary enormously, yet the ratio of the two varies very little, and the results justify, I think, Lenard's conclusion that the distance through which these rays travel only depends on the density of the substance —that is, the mass of matter per unit volume, and not upon the nature of the matter.

These numbers raise a question which I have not yet touched upon, and that is the size of the carriers of the electric charge. Are they or are they not the dimensions of ordinary matter?

We see from Lenard's table that a cathode ray can travel through air at atmospheric pressure a distance of about half a centimetre before the brightness of the phosphorescence falls to about one-half of its original value. Now the mean free path of the molecule of air at this pressure is about 10^{-5} cm., and if a molecule of air were projected it would lose half its momentum in a space comparable with the mean free path. Even if we suppose that it is not the same molecule that is carried, the effect of the obliquity of the collisions would reduce the momentum to one-half in a short multiple of that path.

Thus, from Lenard's experiments on the absorption of the rays outside the tube, it follows on the hypothesis that the cathode rays are charged particles moving with high velocities, that the size of the carriers must be small compared with the dimensions of ordinary atoms or molecules. The assumption of a state of matter more finely subdivided than the atom of an element is a somewhat startling one; but a hypothesis that would involve somewhat similar consequences —viz. that the so-called elements are compounds of some primordial element—has been put forward from time to time by various chemists. Thus, Prout believed that the atoms of all the elements were built up of atoms of hydrogen, and Mr. Norman Lockyer has advanced weighty arguments, founded on spectroscopic consideration, in favour of the composite nature of the elements.

Let us trace the consequence of supposing that the atoms of the elements are aggregations of very small particles, all similar to each other; we shall call such particles corpuscles, so that the atoms of the ordinary elements are made up of corpuscles and holes, the holes being predominant. Let us suppose that at the cathode some of the molecules of the gas get split up into these corpuscles, and that these, charged with negative electricity and moving at a high velocity, form the cathode rays. The distance these rays would travel before losing a given fraction of their momentum would be proportional to the mean free path of the corpuscles. Now, the things these corpuscles strike against are other corpuscles, and not against the molecules as a whole; they are supposed to be able to thread their way between the interstices in the molecule. Thus the mean free path would be proportional to the number of these corpuscles; and, therefore, since each corpuscle has the same mass to the mass of unit volume—that is, to the density of the substance, whatever be its chemical nature or physical state. Thus the mean free path, and therefore the coefficient of absorption, would depend only on the density; this is precisely Lenard's result.

We see, too, on this hypothesis, why the magnetic deflection is the same inside the tube whatever be the nature of the gas, for the carriers of the charge are the corpuscles, and these are the same whatever gas be used. All the carriers may not be reduced to their lowest dimensions; some may be aggregates of two or more corpuscles; these would be differently deflected from the single corpuscle, thus we should get the magnetic spectrum.

I have endeavoured by the following method to get a measurement of the ratio of the mass of these corpuscles to the charge carried by them. A double cylinder with slits in it, such as that used in a former experiment, was placed in front of a cathode which was curved so as to focus to some extent the cathode rays on the slit; behind the slit, in the inner cylinder, a thermal junction was placed which covered the opening so that all the rays which entered the slit struck against the junction, the junction got heated, and knowing the thermal capacity of the junction, we could get the mechanical equivalent of the heat communicated to it. The deflection of the electrometer gave the charge which entered the cylinder.

Thus, if there are N particles entering the cylinder each with a charge e, and Q is the charge inside the cylinder,

$$\mathrm{N}e = \mathrm{Q}.$$

The kinetic energy of these

$$\tfrac{1}{2}\mathrm{N}mv^2 = \mathrm{W}$$

where W is the mechanical equivalent of the heat given to the thermal junction. By measuring the curvature of the rays for a magnetic field, we get

$$\frac{m}{e}v = \mathrm{I}.$$

Thus

$$\frac{m}{e} = \tfrac{1}{2}\frac{\mathrm{Q}\,\mathrm{I}^2}{\mathrm{W}}.$$

In an experiment made at a very low pressure, when the rays were kept on for about one second, the charge was sufficient to raise a capacity of 1·5 microfarads to a potential of 16 volts. Thus

$$\mathrm{Q} = 2{\cdot}4 \times 10^{-6}.$$

The temperature of the thermo junction, whose thermal capacity was 0·005 was raised 3·3° C. by the impact of the rays, thus

$$\begin{aligned}\mathrm{W} &= 3{\cdot}3 \times 0{\cdot}005 \times 4{\cdot}2 \times 10^7 \\ &= 6{\cdot}3 \times 10^5.\end{aligned}$$

The value of I was 280, thus

$$\frac{m}{e} = 1{\cdot}6 \times 10^{-7}.$$

This is very small compared with the value 10^{-4} for the ratio of the mass of an atom of hydrogen to the charge carried by it. If the result stood by itself we might think that it was probable that e was greater than the atomic charge of atom rather than that m was less than the mass of a hydrogen atom. Taken, however, in conjunction with Lenard's results for the absorption of the cathode rays, these numbers seem to favour the hypothesis that the carriers of the charges are smaller than the atoms of hydrogen.

It is interesting to notice that the value of e/m, which we have found from the cathode rays, is of the same order as the value 10^{-7} deduced by Zeeman from his experiments on the effect of a magnetic field on the period of the sodium light.

[J. J. T.]

LONDON: PRINTED BY WILLIAM CLOWES AND SONS, LIMITED,
STAMFORD STREET AND CHARING CROSS.

charge and acquires a negative one; while if the initial charge is a negative one, the cylinder will leak if the initial negative potential is numerically greater than the equilibrium value.

Deflexion of the Cathode Rays by an Electrostatic Field.

An objection very generally urged against the view that the cathode rays are negatively electrified particles, is that hitherto no deflexion of the rays has been observed under a small electrostatic force, and though the rays are deflected when they pass near electrodes connected with sources of large differences of potential, such as induction-coils or electrical machines, the deflexion in this case is regarded by the supporters of the ætherial theory as due to the discharge passing between the electrodes, and not primarily to the electrostatic field. Hertz made the rays travel between two parallel plates of metal placed inside the discharge-tube, but found that they were not deflected when the plates were connected with a battery of storage-cells; on repeating this experiment I at first got the same result, but subsequent experiments showed that the absence of deflexion is due to the conductivity conferred on the rarefied gas by the cathode rays. On measuring this conductivity it was found that it diminished very rapidly as the exhaustion increased; it seemed then that on trying Hertz's experiment at very high exhaustions there might be a chance of detecting the deflexion of the cathode rays by an electrostatic force.

The apparatus used is represented in fig. 2.

Fig. 2.

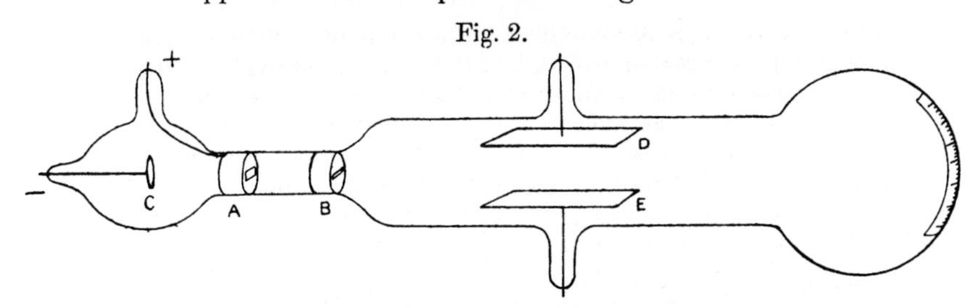

The rays from the cathode C pass through a slit in the anode A, which is a metal plug fitting tightly into the tube and connected with the earth; after passing through a second slit in another earth-connected metal plug B, they travel between two parallel aluminium plates about 5 cm. long by 2 broad and at a distance of 1·5 cm. apart; they then fall on the end of the tube and produce a narrow well-defined phosphorescent patch. A scale pasted on the outside of the tube serves to measure the deflexion of this patch.

At high exhaustions the rays were deflected when the two aluminium plates were connected with the terminals of a battery of small storage-cells; the rays were depressed when the upper plate was connected with the negative pole of the battery, the lower with the positive, and raised when the upper plate was connected with the positive, the lower with the negative pole. The deflexion was proportional to the difference of potential between the plates, and I could detect the deflexion when the potential-difference was as small as two volts. It was only when the vacuum was a good one that the deflexion took place, but that the absence of deflexion is due to the conductivity of the medium is shown by what takes place when the vacuum has just arrived at the stage at which the deflexion begins. At this stage there is a deflexion of the rays when the plates are first connected with the terminals of the battery, but if this connexion is maintained the patch of phosphorescence gradually creeps back to its undeflected position. This is just what would happen if the space between the plates were a conductor, though a very bad one, for then the positive and negative ions between the plates would slowly diffuse, until the positive plate became coated with negative ions, the negative plate with positive ones; thus the electric intensity between the plates would vanish and the cathode rays be free from electrostatic force. Another illustration of this is afforded by what happens when the pressure is low enough to show the deflexion and a large difference of potential, say 200 volts, is established between the plates; under these circumstances there is a large deflexion of the cathode rays, but the medium under the large electromotive force breaks down every now and then and a bright discharge passes between the plates; when this occurs the phosphorescent patch produced by the cathode rays jumps back to its undeflected position. When the cathode rays are deflected by the electrostatic field, the phosphorescent band breaks up into several bright bands separated by comparatively dark spaces; the phenomena are exactly analogous to those observed by Birkeland when the cathode rays are deflected by a magnet, and called by him the magnetic spectrum.

A series of measurements of the deflexion of the rays by the electrostatic force under various circumstances will be found later on in the part of the paper which deals with the velocity of the rays and the ratio of the mass of the electrified particles to the charge carried by them. It may, however, be mentioned here that the deflexion gets smaller as the pressure diminishes, and when in consequence the potential-difference in the tube in the neighbourhood of the cathode increases.

Conductivity of a Gas through which Cathode Rays are passing.

The conductivity of the gas was investigated by means of the apparatus shown in fig. 2. The upper plate D was connected with one terminal of a battery of small storage-cells, the other terminal of which was connected with the earth; the other plate E was connected with one of the coatings of a condenser of one microfarad capacity, the other coating of which was to earth; one pair of quadrants of an electrometer was also connected with E, the other pair of quadrants being to earth. When the cathode rays are passing between the plates the two pairs of quadrants of the electrometer are first connected with each other, and then the connexion between them was broken. If the space between the plates were a non-conductor, the potential of the pair of quadrants not connected with the earth would remain zero and the needle of the electrometer would not move; if, however, the space between the plates were a conductor, then the potential of the lower plate would approach that of the upper, and the needle of the electrometer would be deflected. There is always a deflexion of the electrometer, showing that a current passes between the plates. The magnitude of the current depends very greatly upon the pressure of the gas; so much so, indeed, that it is difficult to obtain consistent readings in consequence of the changes which always occur in the pressure when the discharge passes through the tube.

We shall first take the case when the pressure is only just low enough to allow the phosphorescent patch to appear at the end of the tube; in this case the relation between the current between the plates and the initial difference of potential is represented by the curve shown in fig. 3. In this

Fig. 3.

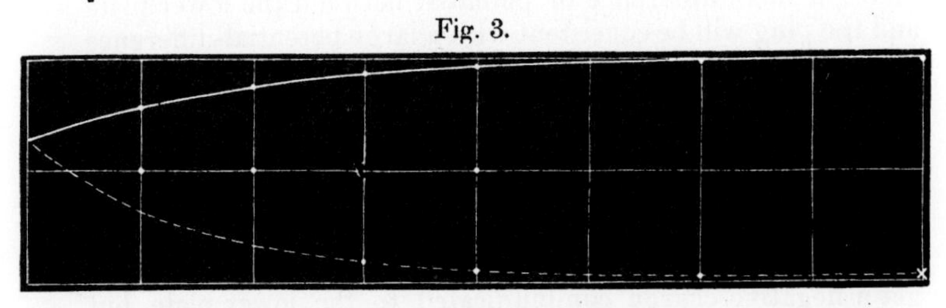

figure the abscissæ represent the initial difference of potential between the plates, each division representing two volts, and the ordinates the rise in potential of the lower plate in one minute each division again representing two volts. The quantity of electricity which has passed between the plates in one minute is the quantity required to raise 1 microfarad to the potential-difference shown by the curve. The upper and lower curve relates to the case when the upper plate is connected with the negative and positive pole respectively of the battery.

Even when there is no initial difference of potential between the plates the lower plate acquires a negative charge from the impact on it of some of the cathode rays.

We see from the curve that the current between the plates soon reaches a value where it is only slightly affected by an increase in the potential-difference between the plates; this is a feature common to conduction through gases traversed by Röntgen rays, by uranium rays, by ultra-violet light, and, as we now see, by cathode rays. The rate of leak is not greatly different whether the upper plate be initially positively or negatively electrified.

The current between the plates only lasts for a short time; it ceases long before the potential of the lower plate approaches that of the upper. Thus, for example, when the potential of the upper plate was about 400 volts above that of the earth, the potential of the lower plate never rose above 6 volts: similarly, if the upper plate were connected with the negative pole of the battery, the fall in potential of the lower plate was very small in comparison with the potential-difference between the upper plate and the earth.

These results are what we should expect if the gas between the plates and the plug B (fig. 2) were a very much better conductor than the gas between the plates, for the lower plate will be in a steady state when the current coming to it from the upper plate is equal to the current going from it to the plug: now if the conductivity of the gas between the plate and the plug is much greater than that between the plates, a small difference of potential between the lower plate and the plug will be consistent with a large potential-difference between the plates.

So far we have been considering the case when the pressure is as high as is consistent with the cathode rays reaching the end of the tube; we shall now go to the other extreme and consider the case when the pressure is as low as is consistent with the passage of a discharge through the bulb. In this case, when the plates are not connected with the battery we get a negative charge communicated to the lower plate, but only very slowly in comparison with the effect in the previous case. When the upper plate is connected with the negative pole of a battery, this current to the lower plate is only slightly increased even when the difference of potential is as much as 400 volts: a small potential-difference of about

20 volts seems slightly to decrease the rate of leak. Potential-differences much exceeding 400 volts cannot be used, as though the dielectric between the plates is able to sustain them for some little time, yet after a time an intensely bright arc flashes across between the plates and liberates so much gas as to spoil the vacuum. The lines in the spectrum of this glare are chiefly mercury lines; its passage leaves very peculiar markings on the aluminium plates.

If the upper plate was charged positively, then the negative charge communicated to the lower plate was diminished, and stopped when the potential-difference between the plates was about 20 volts; but at the lowest pressure, however great (up to 400 volts) the potential-difference, there was no leak of positive electricity to the lower plate at all comparable with the leak of negative electricity to this plate when the two plates were disconnected from the battery. In fact at this very low pressure all the facts are consistent with the view that the effects are due to the negatively electrified particles travelling along the cathode rays, the rest of the gas possessing little conductivity. Some experiments were made with a tube similar to that shown in fig. 2, with the exception that the second plug B was absent, so that a much greater number of cathode rays passed between the plates. When the upper plate was connected with the positive pole of the battery a luminous discharge with well-marked striations passed between the upper plate and the earth-connected plug through which the cathode rays were streaming; this occurred even though the potential-difference between the plate and the plug did not exceed 20 volts. Thus it seems that if we supply cathode rays from an external source to the cathode a small potential-difference is sufficient to produce the characteristic discharge through a gas.

Magnetic Deflexion of the Cathode Rays in Different Gases.

The deflexion of the cathode rays by the magnetic field was studied with the aid of the apparatus shown in fig. 4. The cathode was placed in a side-tube fastened on to a bell-jar; the opening between this tube and the bell-jar was closed by a metallic plug with a slit in it; this plug was connected with the earth and was used as the anode. The cathode rays passed through the slit in this plug into the bell-jar, passing in front of a vertical plate of glass ruled into small squares. The bell-jar was placed between two large parallel coils arranged as a Helmholtz galvanometer. The course of the rays was determined by taking photographs of the bell-jar when the cathode rays were passing through it; the divisions on the plate enabled the path of the rays to be determined. Under the action of the magnetic field the narrow beam of cathode rays spreads out into a broad fan-shaped luminosity in the gas. The luminosity in this fan is not uniformly

Fig. 4.

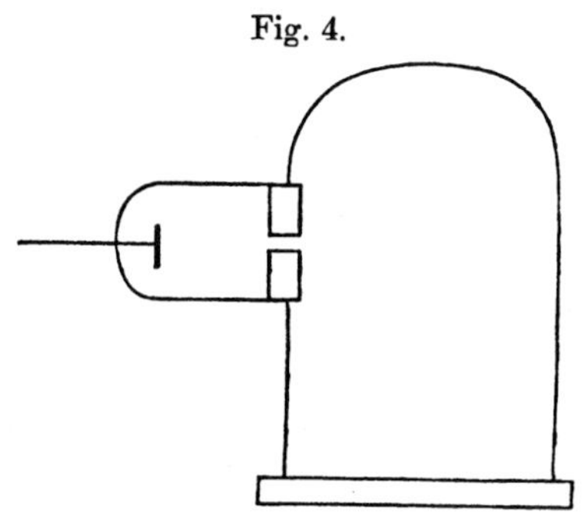

distributed, but is condensed along certain lines. The phosphorescence on the glass is also not uniformly distributed; it is much spread out, showing that the beam consists of rays which are not all deflected to the same extent by the magnet. The luminosity on the glass is crossed by bands along which the luminosity is very much greater than in the adjacent parts. These bright and dark bands are called by Birkeland, who first observed them, the magnetic spectrum. The brightest spots on the glass are by no means always the terminations of the brightest streaks of luminosity in the gas; in fact, in some cases a very bright spot on the glass is not connected with the cathode by any appreciable luminosity, though there may be plenty of luminosity in other parts of the gas. One very interesting point brought out by the photographs is that in a given magnetic field, and with a given mean potential-difference between the terminals, the path of the rays is independent of the nature of the gas. Photographs were taken of the discharge in hydrogen, air, carbonic acid, methyl iodide, *i. e.*, in gases whose densities range from 1 to 70, and yet, not only were the paths of the most deflected rays the same in all cases, but even the details, such as the distribution of the bright and dark spaces, were the same; in fact, the photographs could hardly be distinguished from each other. It is to be ncted that the pressures were not the same; the pressures in the different gases were adjusted so that the mean potential-differences between the cathode and the anode were the same in all the gases. When the pressure of a gas is lowered, the potential-difference between the terminals increases, and the deflexion of the rays produced by a magnet diminishes, or at any rate the deflexion of the rays when the phosphorescence is a maximum diminishes. If an air-break is inserted an effect of the same kind is produced.

In the experiments with different gases, the pressures were as high as was consistent with the appearance of the phosphorescence on the glass, so as to ensure having as much as possible of the gas under consideration in the tube.

As the cathode rays carry a charge of negative electricity, are deflected by an electrostatic force as if they were negatively electrified, and are acted on by a magnetic force in just the way in which this force would act on a negatively electrified body moving along the path of these rays, I can see no escape from the conclusion that they are charges of negative electricity carried by particles of matter. The question next arises, What are these particles? are they atoms, or molecules, or matter in a still finer state of subdivision? To throw some light on this point, I have made a series of measurements of the ratio of the mass of these particles to the charge carried by it. To determine this quantity, I have used two independent methods. The first of these is as follows:—Suppose we consider a bundle of homogeneous cathode rays. Let m be the mass of each of the particles, e the charge carried by it. Let N be the number of particles passing across any section of the beam in a given time; then Q the quantity of electricity carried by these particles is given by the equation

$$Ne = Q.$$

We can measure Q if we receive the cathode rays in the inside of a vessel connected with an electrometer. When these rays strike against a solid body, the temperature of the body is raised; the kinetic energy of the moving particles being converted into heat; if we suppose that all this energy is converted into heat, then if we measure the increase in the temperature of a body of known thermal capacity caused by the impact of these rays, we can determine W, the kinetic energy of the particles, and if v is the velocity of the particles,

$$\tfrac{1}{2}Nmv^2 = W.$$

If ρ is the radius of curvature of the path of these rays in a uniform magnetic field H, then

$$\frac{mv}{e} = H\rho = I,$$

where I is written for $H\rho$ for the sake of brevity. From these equations we get

$$\frac{1}{2}\frac{m}{e}v^2 = \frac{W}{Q}.$$

$$v = \frac{2W}{QI},$$

$$\frac{m}{e} = \frac{I^2Q}{2W}.$$

Thus, if we know the values of Q, W, and I, we can deduce the values of v and m/e.

To measure these quantities, I have used tubes of three different types. The first I tried is like that represented in fig. 2, except that the plates E and D are absent, and two coaxial cylinders are fastened to the end of the tube. The rays from the cathode C fall on the metal plug B, which is connected with the earth, and serves for the anode; a horizontal slit is cut in this plug. The cathode rays pass through this slit, and then strike against the two coaxial cylinders at the end of the tube; slits are cut in these cylinders, so that the cathode rays pass into the inside of the inner cylinder. The outer cylinder is connected with the earth, the inner cylinder, which is insulated from the outer one, is connected with an electrometer, the deflexion of which measures Q, the quantity of electricity brought into the inner cylinder by the rays. A thermo-electric couple is placed behind the slit in the inner cylinder; this couple is made of very thin strips of iron and copper fastened to very fine iron and copper wires. These wires passed through the cylinders, being insulated from them, and through the glass to the outside of the tube, where they were connected with a low-resistance galvanometer, the deflexion of which gave data for calculating the rise of temperature of the junction produced by the impact against it of the cathode rays. The strips of iron and copper were large enough to ensure that every cathode ray which entered the inner cylinder struck against the junction. In some of the tubes the strips of iron and copper were placed end to end, so that some of the rays struck against the iron, and others against the copper; in others, the strip of one metal was placed in front of the other; no difference, however, could be detected between the results got with these two arrangements. The strips of iron and copper were weighed, and the thermal capacity of the junction calculated. In one set of junctions this capacity was 5×10^{-3}, in another 3×10^{-3}. If we assume that the cathode rays which strike against the junction give their energy up to it, the deflexion of the galvanometer gives us W or $\frac{1}{2}Nmv^2$.

The value of I, *i. e.*, Hρ, where ρ is the curvature of the path of the rays in a magnetic field of strength H was found as follows :—The tube was fixed between two large circular coils placed parallel to each other, and separated by a distance equal to the radius of either; these coils produce a uniform magnetic field, the strength of which is got by measuring with an ammeter the strength of the current passing through them. The cathode rays are thus in a uniform field, so that their path is circular. Suppose that the rays, when deflected by a magnet, strike against the glass of the tube at E

Fig. 5.

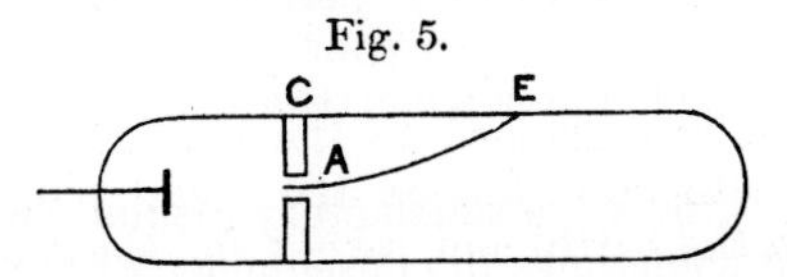

(fig. 5), then, if ρ is the radius of the circular path of the rays,

$$2\rho = \frac{CE^2}{AC} + AC\,;$$

thus, if we measure CE and AC we have the means of determining the radius of curvature of the path of the rays.

The determination of ρ is rendered to some extent uncertain, in consequence of the pencil of rays spreading out under the action of the magnetic field, so that the phosphorescent patch at E is several millimetres long; thus values of ρ differing appreciably from each other will be got by taking E at different points of this phosphorescent patch. Part of this patch was, however, generally considerably brighter than the rest; when this was the case, E was taken as the brightest point; when such a point of maximum brightness did not exist, the middle of the patch was taken for E. The uncertainty in the value of ρ thus introduced amounted sometimes to about 20 per cent.; by this I mean that if we took E first at one extremity of the patch and then at the other, we should get values of ρ differing by this amount.

The measurement of Q, the quantity of electricity which enters the inner cylinder, is complicated by the cathode rays making the gas through which they pass a conductor, so that though the insulation of the inner cylinder was perfect when the rays were off, it was not so when they were passing through the space between the cylinders; this caused some of the charge communicated to the inner cylinder to leak away so that the actual charge given to the cylinder by the cathode rays was larger than that indicated by the electrometer.

To make the error from this cause as small as possible, the inner cylinder was connected to the largest capacity available, 1·5 microfarad, and the rays were only kept on for a short time, about 1 or 2 seconds, so that the alteration in potential of the inner cylinder was not large, ranging in the various experiments from about ·5 to 5 volts. Another reason why it is necessary to limit the duration of the rays to as short a time as possible, is to avoid the correction for the loss of heat from the thermo-electric junction by conduction along the wires; the rise in temperature of the junction was of the order 2° C.; a series of experiments showed that with the same tube and the same gaseous pressure Q and W were proportional to each other when the rays were not kept on too long.

Tubes of this kind gave satisfactory results, the chief drawback being that sometimes in consequence of the charging up of the glass of the tube, a secondary discharge started from the cylinder to the walls of the tube, and the cylinders were surrounded by glow; when this glow appeared, the readings were very irregular; the glow could, however, be got rid of by pumping and letting the tube rest for some time. The results got with this tube are given in the Table under the heading Tube 1.

The second type of tube was like that used for photographing the path of the rays (fig. 4); double cylinders with a thermo-electric junction like those used in the previous tube were placed in the line of fire of the rays, the inside of the bell-jar was lined with copper gauze connected with the earth. This tube gave very satisfactory results; we were never troubled with any glow round the cylinders, and the readings were most concordant; the only drawback was that as some of the connexions had to be made with sealing-wax, it was not possible to get the highest exhaustions with this tube, so that the range of pressure for this tube is less than that for tube 1. The results got with this tube are given in the Table under the heading Tube 2.

The third type of tube was similar to the first, except that the openings in the two cylinders were made very much smaller; in this tube the slits in the cylinders were replaced by small holes, about 1·5 millim. in diameter. In consequence of the smallness of the openings, the magnitude of the effects was very much reduced; in order to get measurable results it was necessary to reduce the capacity of the condenser in connexion with the inner cylinder to ·15 microfarad, and to make the galvanometer exceedingly sensitive, as the rise in temperature of the thermo-electric junction was in these experiments only about ·5° C. on the average. The results obtained in this tube are given in the Table under the heading Tube 3.

The results of a series of measurements with these tubes are given in the following Table :—

Gas.	Value of W/Q.	I.	m/e.	v.
Tube 1.				
Air	$4{\cdot}6 \times 10^{11}$	230	$\cdot 57 \times 10^{-7}$	4×10^{9}
Air	$1{\cdot}8 \times 10^{12}$	350	$\cdot 34 \times 10^{-7}$	1×10^{10}
Air	$6{\cdot}1 \times 10^{11}$	230	$\cdot 43 \times 10^{-7}$	$5{\cdot}4 \times 10^{9}$
Air	$2{\cdot}5 \times 10^{12}$	400	$\cdot 32 \times 10^{-7}$	$1{\cdot}2 \times 10^{10}$
Air	$5{\cdot}5 \times 10^{11}$	230	$\cdot 48 \times 10^{-7}$	$4{\cdot}8 \times 10^{9}$
Air	1×10^{12}	285	$\cdot 4 \times 10^{-7}$	7×10^{9}
Air	1×10^{12}	285	$\cdot 4 \times 10^{-7}$	7×10^{9}
Hydrogen	6×10^{12}	205	$\cdot 35 \times 10^{-7}$	6×10^{9}
Hydrogen	$2{\cdot}1 \times 10^{12}$	460	$\cdot 5 \times 10^{-7}$	$9{\cdot}2 \times 10^{9}$
Carbonic acid	$8{\cdot}4 \times 10^{11}$	260	$\cdot 4 \times 10^{-7}$	$7{\cdot}5 \times 10^{9}$
Carbonic acid	$1{\cdot}47 \times 10^{12}$	340	$\cdot 4 \times 10^{-7}$	$8{\cdot}5 \times 10^{9}$
Carbonic acid	$3{\cdot}0 \times 10^{12}$	480	$\cdot 39 \times 10^{-7}$	$1{\cdot}3 \times 10^{10}$
Tube 2.				
Air	$2{\cdot}8 \times 10^{11}$	175	$\cdot 53 \times 10^{-7}$	$3{\cdot}3 \times 10^{9}$
Air	$4{\cdot}4 \times 10^{11}$	195	$\cdot 47 \times 10^{-7}$	$4{\cdot}1 \times 10^{9}$
Air	$3{\cdot}5 \times 10^{11}$	181	$\cdot 47 \times 10^{-7}$	$3{\cdot}8 \times 10^{9}$
Hydrogen	$2{\cdot}8 \times 10^{11}$	175	$\cdot 53 \times 10^{-7}$	$3{\cdot}3 \times 10^{9}$
Air	$2{\cdot}5 \times 10^{11}$	160	$\cdot 51 \times 10^{-7}$	$3{\cdot}1 \times 10^{9}$
Carbonic acid	2×10^{11}	148	$\cdot 54 \times 10^{-7}$	$2{\cdot}5 \times 10^{9}$
Air	$1{\cdot}8 \times 10^{11}$	151	$\cdot 63 \times 10^{-7}$	$2{\cdot}3 \times 10^{9}$
Hydrogen	$2{\cdot}8 \times 10^{11}$	175	$\cdot 53 \times 10^{-7}$	$3{\cdot}3 \times 10^{9}$
Hydrogen	$4{\cdot}4 \times 10^{11}$	201	$\cdot 46 \times 10^{-7}$	$4{\cdot}4 \times 10^{9}$
Air	$2{\cdot}5 \times 10^{11}$	176	$\cdot 61 \times 10^{-7}$	$2{\cdot}8 \times 10^{9}$
Air	$4{\cdot}2 \times 10^{11}$	200	$\cdot 48 \times 10^{-7}$	$4{\cdot}1 \times 10^{9}$
Tube 3.				
Air	$2{\cdot}5 \times 10^{11}$	220	$\cdot 9 \times 10^{-7}$	$2{\cdot}4 \times 10^{9}$
Air	$3{\cdot}5 \times 10^{11}$	225	$\cdot 7 \times 10^{-7}$	$3{\cdot}2 \times 10^{9}$
Hydrogen	3×10^{11}	250	$1{\cdot}0 \times 10^{-7}$	$2{\cdot}5 \times 10^{9}$

It will be noticed that the value of m/e is considerably greater for Tube 3, where the opening is a small hole, than for Tubes 1 and 2, where the opening is a slit of much greater area. I am of opinion that the values of m/e got from Tubes 1 and 2 are too small, in consequence of the leakage from the inner cylinder to the outer by the gas being rendered a conductor by the passage of the cathode rays.

It will be seen from these tables that the value of m/e is independent of the nature of the gas. Thus, for the first tube the mean for air is $\cdot 40 \times 10^{-7}$, for hydrogen $\cdot 42 \times 10^{-7}$, and for carbonic acid gas $\cdot 4 \times 10^{-7}$; for the second tube the mean for air is $\cdot 52 \times 10^{-7}$, for hydrogen $\cdot 50 \times 10^{-7}$, and for carbonic acid gas $\cdot 54 \times 10^{-7}$.

Experiments were tried with electrodes made of iron instead of aluminium; this altered the appearance of the discharge and the value of v at the same pressure, the values of m/e were, however, the same in the two tubes; the effect produced by different metals on the appearance of the discharge will be described later on.

In all the preceding experiments, the cathode rays were first deflected from the cylinder by a magnet, and it was then found that there was no deflexion either of the electrometer or the galvanometer, so that the deflexions observed were entirely due to the cathode rays; when the glow mentioned previously surrounded the cylinders there was a deflexion of the electrometer even when the cathode rays were deflected from the cylinder.

Before proceeding to discuss the results of these measurements I shall describe another method of measuring the quantities m/e and v of an entirely different kind from the preceding; this method is based upon the deflexion of the cathode rays in an electrostatic field. If we measure the deflexion experienced by the rays when traversing a given length under a uniform electric intensity, and the deflexion of the rays when they traverse a given distance under a uniform magnetic field, we can find the values of m/e and v in the following way:—

Let the space passed over by the rays under a uniform electric intensity F be l, the time taken for the rays to traverse this space is l/v, the velocity in the direction of F is therefore

$$\frac{Fe}{m}\frac{l}{v},$$

so that θ, the angle through which the rays are deflected when they leave the electric field and enter a region free from electric force, is given by the equation

$$\theta = \frac{Fe}{m} \frac{l}{v^2}.$$

If, instead of the electric intensity, the rays are acted on by a magnetic force H at right angles to the rays, and extending across the distance l, the velocity at right angles to the original path of the rays is

$$\frac{Hev}{m} \frac{l}{v},$$

so that ϕ, the angle through which the rays are deflected when they leave the magnetic field, is given by the equation

$$\phi = \frac{He}{m} \frac{l}{v}.$$

From these equations we get

$$v = \frac{\phi}{\theta} \frac{F}{H}$$

and

$$\frac{m}{e} = \frac{H^2\theta \, . \, l}{F\phi^2}.$$

In the actual experiments H was adjusted so that $\phi = \theta$; in this case the equations become

$$v = \frac{F}{H},$$

$$\frac{m}{e} = \frac{H^2 l}{F\theta}.$$

The apparatus used to measure v and m/e by this means is that represented in fig. 2. The electric field was produced by connecting the two aluminium plates to the terminals of a battery of storage-cells. The phosphorescent patch at the end of the tube was deflected, and the deflexion measured by a scale pasted to the end of the tube. As it was necessary to darken the room to see the phosphorescent patch, a needle coated with luminous paint was placed so that by a screw it could be moved up and down the scale; this needle could be seen when the room was darkened, and it was moved until it coincided with the phosphorescent patch. Thus, when light was admitted, the deflexion of the phosphorescent patch could be measured.

The magnetic field was produced by placing outside the tube two coils whose diameter was equal to the length of the plates; the coils were placed so that they covered the space occupied by the plates, the distance between the coils was equal to the radius of either. The mean value of the magnetic force over the length l was determined in the following way: a narrow coil C whose length was l, connected with a ballistic galvanometer, was placed between the coils; the plane of the windings of C was parallel to the planes of the coils; the cross section of the coil was a rectangle 5 cm. by 1 cm. A given current was sent through the outer coils and the kick α of the galvanometer observed when this current was reversed. The coil C was then placed at the centre of two very large coils, so as to be in a field of uniform magnetic force: the current through the large coils was reversed and the kick β of the galvanometer again observed; by comparing α and β we can get the mean value of the magnetic force over a length l; this was found to be

$$60 \times \iota,$$

where ι is the current flowing through the coils.

A series of experiments was made to see if the electrostatic deflexion was proportional to the electric intensity between the plates; this was found to be the case. In the following experiments the current through the coils was adjusted so that the electrostatic deflexion was the same as the magnetic:—

Gas.	θ.	H.	F.	l.	m/e.	v.
Air	8/110	5·5	$1·5 \times 10^{10}$	5	$1·3 \times 10^{-7}$	$2·8 \times 10^{9}$
Air	9·5/110	5·4	$1·5 \times 10^{10}$	5	$1·1 \times 10^{-7}$	$2·8 \times 10^{9}$
Air	13/110	6·6	$1·5 \times 10^{10}$	5	$1·2 \times 10^{-7}$	$2·3 \times 10^{9}$
Hydrogen	9/110	6·3	$1·5 \times 10^{10}$	5	$1·5 \times 10^{-7}$	$2·5 \times 10^{9}$
Carbonic acid...	11/110	6·9	$1·5 \times 10^{10}$	5	$1·5 \times 10^{-7}$	$2·2 \times 10^{9}$
Air	6/110	5	$1·8 \times 10^{10}$	5	$1·3 \times 10^{-7}$	$3·6 \times 10^{9}$
Air	7/110	3·6	1×10^{10}	5	$1·1 \times 10^{-7}$	$2·8 \times 10^{9}$

The cathode in the first five experiments was aluminium, in the last two experiments it was made of platinum; in the last experiment Sir William Crookes's method of getting rid of the mercury vapour by inserting tubes of pounded sulphur, sulphur iodide, and copper filings between the bulb and the pump was adopted. In the calculation of m/e and v no allowance has been made for the magnetic force due to the coil in the region outside the plates; in this region the magnetic force will be in the opposite direction to that between the plates, and will tend to bend the cathode rays in the opposite direction: thus the effective value of H will be smaller than the value used in the equations, so that the values of m/e are larger, and those of v less than they would be if this correction were applied. This method of determining the values of m/e and v is much less laborious and probably more accurate than the former method; it cannot, however, be used over so wide a range of pressures.

From these determinations we see that the value of m/e is independent of the nature of the gas, and that its value 10^{-7} is very small compared with the value 10^{-4}, which is the smallest value of this quantity previously known, and which is the value for the hydrogen ion in electrolysis.

Thus for the carriers of the electricity in the cathode rays m/e is very small compared with its value in electrolysis. The smallness of m/e may be due to the smallness of m or the largeness of e, or to a combination of these two. That the carriers of the charges in the cathode rays are small compared with ordinary molecules is shown, I think, by Lenard's results as to the rate at which the brightness of the phosphorescence produced by these rays diminishes with the length of path travelled by the ray. If we regard this phosphorescence as due to the impact of the charged particles, the distance through which the rays must travel before the phosphorescence fades to a given fraction (say $1/e$, where $e=2{\cdot}71$) of its original intensity, will be some moderate multiple of the mean free path. Now Lenard found that this distance depends solely upon the density of the medium, and not upon its chemical nature or physical state. In air at atmospheric pressure the distance was about half a centimetre, and this must be comparable with the mean free path of the carriers through air at atmospheric pressure. But the mean free path of the molecules of air is a quantity of quite a different order. The carrier, then, must be small compared with ordinary molecules.

The two fundamental points about these carriers seem to me to be (1) that these carriers are the same whatever the gas through which the discharge passes, (2) that the mean free paths depend upon nothing but the density of the medium traversed by these rays.

It might be supposed that the independence of the mass of the carriers of the gas through which the discharge passes was due to the mass concerned being the quasi mass which a charged body possesses in virtue of the electric field set up in its neighbourhood; moving the body involves the production of a varying electric field, and, therefore, of a certain amount of energy which is proportional to the square of the velocity. This causes the charged body to behave as if its mass were increased by a quantity, which for a charged sphere is $\frac{1}{5}e^2/\mu a$ ('Recent Researches in Electricity and Magnetism'), where e is the charge and a the radius of the sphere. If we assume that it is this mass which we are concerned with in the cathode rays, since m/e would vary as e/a, it affords no clue to the explanation of either of the properties (1 and 2) of these rays. This is not by any means the only objection to this hypothesis, which I only mention to show that it has not been overlooked.

The explanation which seems to me to account in the most simple and straightforward manner for the facts is founded on a view of the constitution of the chemical elements which has been favourably entertained by many chemists: this view is that the atoms of the different chemical elements are different aggregations of atoms of the same kind. In the form in which this hypothesis was enunciated by Prout, the atoms of the different elements were hydrogen atoms; in this precise form the hypothesis is not tenable, but if we substitute for hydrogen some unknown primordial substance X, there is nothing known which is inconsistent with this hypothesis, which is one that has been recently supported by Sir Norman Lockyer for reasons derived from the study of the stellar spectra.

If, in the very intense electric field in the neighbourhood of the cathode, the molecules of the gas are dissociated and are split up, not into the ordinary chemical atoms, but into these primordial atoms, which we shall for brevity call corpuscles; and if these corpuscles are charged with electricity and projected from the cathode by the electric field, they would behave exactly like the cathode rays. They would evidently give a value of m/e which is independent of the nature of the gas and its pressure, for the carriers are the same whatever the gas may be; again, the mean free paths of these corpuscles would depend solely upon the density of the medium through which they pass. For the molecules of the medium are composed of a number of such corpuscles separated by considerable spaces; now the collision between a single corpuscle and the molecule will not be between the corpuscles and the molecule as a whole, but between this corpuscle and the individual corpuscles which form the molecule; thus the number of collisions the particle makes as it moves through a crowd of these molecules will be proportional, not to the number of the

Z 2

molecules in the crowd, but to the number of the individual corpuscles. The mean free path is inversely proportional to the number of collisions in unit time, and so is inversely proportional to the number of corpuscles in unit volume; now as these corpuscles are all of the same mass, the number of corpuscles in unit volume will be proportional to the mass of unit volume, that is the mean free path will be inversely proportional to the density of the gas. We see, too, that so long as the distance between neighbouring corpuscles is large compared with the linear dimensions of a corpuscle the mean free path will be independent of the way they are arranged, provided the number in unit volume remains constant, that is the mean free path will depend only on the density of the medium traversed by the corpuscles, and will be independent of its chemical nature and physical state: this from Lenard's very remarkable measurements of the absorption of the cathode rays by various media, must be a property possessed by the carriers of the charges in the cathode rays.

Thus on this view we have in the cathode rays matter in a new state, a state in which the subdivision of matter is carried very much further than in the ordinary gaseous state: a state in which all matter—that is, matter derived from different sources such as hydrogen, oxygen, &c.—is of one and the same kind; this matter being the substance from which all the chemical elements are built up.

With appliances of ordinary magnitude, the quantity of matter produced by means of the dissociation at the cathode is so small as to almost to preclude the possibility of any direct chemical investigation of its properties. Thus the coil I used would, I calculate, if kept going uninterruptedly night and day for a year, produce only about one three-millionth part of a gramme of this substance.

The smallness of the value of m/e is, I think, due to the largeness of e as well as the smallness of m. There seems to me to be some evidence that the charges carried by the corpuscles in the atom are large compared with those carried by the ions of an electrolyte. In the molecule of HCl, for example, I picture the components of the hydrogen atoms as held together by a great number of tubes of electrostatic force; the components of the chlorine atom are similarly held together, while only one stray tube binds the hydrogen atom to the chlorine atom. The reason for attributing this high charge to the constituents of the atom is derived from the values of the specific inductive capacity of gases: we may imagine that the specific inductive capacity of a gas is due to the setting in the electric field of the electric doublet formed by the two oppositely electrified atoms which form the molecule of the gas. The measurements of the specific inductive capacity show, however, that this is very approximately an additive quantity: that is, that we can assign a certain value to each element, and find the specific inductive capacity of HCl by adding the value for hydrogen to the value for chlorine; the value of H_2O by adding twice the value for hydrogen to the value for oxygen, and so on. Now the electrical moment of the doublet formed by a positive charge on one atom of the molecule and a negative charge on the other atom would not be an additive property; if, however, each atom had a definite electrical moment, and this were large compared with the electrical moment of the two atoms in the molecule, then the electrical moment of any compound, and hence its specific inductive capacity, would be an additive property. For the electrical moment of the atom, however, to be large compared with that of the molecule, the charge on the corpuscles would have to be very large compared with those on the ion.

If we regard the chemical atom as an aggregation of a number of primordial atoms, the problem of finding the configurations of stable equilibrium for a number of equal particles acting on each other according to some law of force —whether that of Boscovich, where the force between them is a repulsion when they are separated by less than a certain critical distance, and an attraction when they are separated by a greater distance, or even the simpler case of a number of mutually repellent particles held together by a central force —is of great interest in connexion with the relation between the properties of an element and its atomic weight. Unfortunately the equations which determine the stability of such a collection of particles increase so rapidly in complexity with the number of particles that a general mathematical investigation is scarcely possible. We can, however, obtain a good deal of insight into the general laws which govern such configurations by the use of models, the simplest of which is the floating magnets of Professor Mayer. In this model the magnets arrange themselves in equilibrium under their mutual repulsions and a central attraction caused by the pole of a large magnet placed above the floating magnets.

A study of the forms taken by these magnets seems to me to be suggestive in relation to the periodic law. Mayer showed that when the number of floating magnets did not exceed 5 they arranged themselves at the corners of a regular polygon— 5 at the corners of a pentagon, 4 at the corners of a square, and so on. When the number exceeds 5, however, this law no longer holds : thus 6 magnets do not arrange themselves at the corners of a hexagon, but divide into two systems, consisting of 1 in the middle surrounded by 5 at the corners of a pentagon. For 8 we have two in the inside and 6 outside ; this arrangement in two systems, an inner and an outer, lasts up to 18 magnets. After this we have three systems : an inner, a middle, and an outer ; for a still larger number of magnets we have four systems, and so on.

Mayer found the arrangement of magnets was as follows:—

1.	2.	3.	4.	5.
1.5	2.6	3.7	4.8	5.9
1.6	2.7	3.8	4.9	
1.7				
1.5.9	2.7.10	3.7.10	4.8.12	5.9.12
1.6.9	2.8.10	3.7.11	4.8.13	5.9.13
1.6.10	2.7.11	3.8.10	4.9.12	
1.6.11		3.8.11	4.9.13	
		3.8.12		
		3.8.13		
1.5. 9.12	2.7.10.15	3.7.12.13	4.9.13.14	
1.5. 9.13	2.7.12.14	3.7.12.14	4.9.13.15	
1.6. 9.12		3.7.13.14	4.9.14.15	
1.6.10.12		3.7.13.15		
1.6.10.13				
1.6.11.12				
1.6.11.13				
1.6.11.14				
1.6.11.15				
1.7.12.14				

where, for example, **1 . 6 . 10 . 12** means an arrangement with one magnet in the middle, then a ring of six, then a ring of ten, and a ring of twelve outside.

Now suppose that a certain property is associated with two magnets forming a group by themselves; we should have this property with 2 magnets, again with 8 and 9, again with 19 and 20, and again with 34, 35, and so on. If we regard the system of magnets as a model of an atom, the number of magnets being proportional to the atomic weight, we should have this property occurring in elements of atomic weight 2, (8, 9), 19, 20, (34, 35). Again, any property conferred by three magnets forming a system by themselves would occur with atomic weights 3, 10, and 11; 20, 21, 22, 23, and 24; 35, 36, 37 and 39; in fact, we should have something quite analogous to the periodic law, the first series corresponding to the arrangement of the magnets in a single group, the second series to the arrangement in two groups, the third series in three groups, and so on.

Table IV.—Selected Fusing and Boiling Points on the Proposed British Association Scale.

Substance.	F.P.	Substance.	B.P.
Tin....................	231 9 °	Aniline	184·1
Bismuth..............	269·2	Naphthalene	218·0
Cadmium	320·7	Benzophenone	305·8
Lead	327·7	Mercury	356·7
Zinc	419·0	Sulphur...............	444·5
Antimony	629·5	Cadmium	756
Aluminium	654·5	Zinc	916

My thanks are due to several Members of the Electrical Standards Committee of the British Association and others, who have kindly revised the proofs of this article.

LVIII. *On the Masses of the Ions in Gases at Low Pressures. By* J. J. Thomson, *M.A., F.R.S., Cavendish Professor of Experimental Physics, Cambridge**.

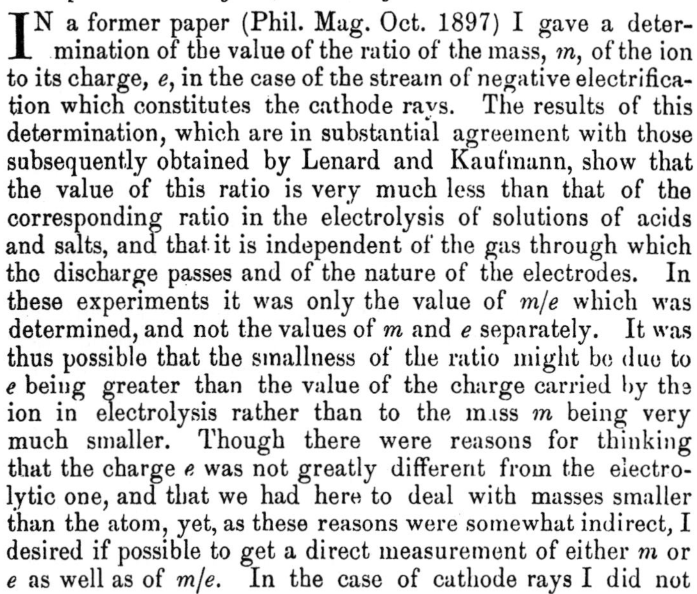

IN a former paper (Phil. Mag. Oct. 1897) I gave a determination of the value of the ratio of the mass, m, of the ion to its charge, e, in the case of the stream of negative electrification which constitutes the cathode rays. The results of this determination, which are in substantial agreement with those subsequently obtained by Lenard and Kaufmann, show that the value of this ratio is very much less than that of the corresponding ratio in the electrolysis of solutions of acids and salts, and that it is independent of the gas through which the discharge passes and of the nature of the electrodes. In these experiments it was only the value of m/e which was determined, and not the values of m and e separately. It was thus possible that the smallness of the ratio might be due to e being greater than the value of the charge carried by the ion in electrolysis rather than to the mass m being very much smaller. Though there were reasons for thinking that the charge e was not greatly different from the electrolytic one, and that we had here to deal with masses smaller than the atom, yet, as these reasons were somewhat indirect, I desired if possible to get a direct measurement of either m or e as well as of m/e. In the case of cathode rays I did not

* Communicated by the Author: read at the Meeting of the British Association at Dover.

see my way to do this; but another case, where negative electricity is carried by charged particles (*i. e.* when a negatively electrified metal plate in a gas at low pressure is illuminated by ultra-violet light), seemed more hopeful, as in this case we can determine the value of e by the method I previously employed to determine the value of the charge carried by the ions produced by Röntgen-ray radiation (Phil. Mag. Dec. 1898). The following paper contains an account of measurements of m/e and e for the negative electrification discharged by ultra-violet light, and also of m/e for the negative electrification produced by an incandescent carbon filament in an atmosphere of hydrogen. I may be allowed to anticipate the description of these experiments by saying that they lead to the result that the value of m/e in the case of the ultra-violet light, and also in that of the carbon filament, is the same as for the cathode rays; and that in the case of the ultra-violet light, e is the same in magnitude as the charge carried by the hydrogen atom in the electrolysis of solutions. In this case, therefore, we have clear proof that the ions have a very much smaller mass than ordinary atoms; so that in the convection of negative electricity at low pressures we have something smaller even than the atom, something which involves the splitting up of the atom, inasmuch as we have taken from it a part, though only a small one, of its mass.

The method of determining the value of m/e for the ions carrying the negative electrification produced by ultra-violet light is as follows:—Elster and Geitel (Wied. *Ann.* xli. p. 166) have shown that the rate of escape of the negative electrification at low pressures is much diminished by magnetic force if the lines of magnetic force are at right angles to the lines of electric force. Let us consider what effect a magnetic force would have on the motion of a negatively electrified particle. Let the electric force be uniform and parallel to the axis of x, while the magnetic force is also uniform and parallel to the axis of z. Let the pressure be so low that the mean free path of the particles is long compared with the distance they move while under observation, so that we may leave out of account the effect of collisions on the movements of the particles.

If m is the mass of a particle, e its charge, X the electric force, H the magnetic force, the equations of motion are:—

$$m\frac{d^2x}{dt^2} = \mathrm{X}e - \mathrm{H}e\frac{dy}{dt},$$

$$m\frac{d^2y}{dt^2} = \mathrm{H}e\frac{dx}{dt}.$$

Eliminating x we have :—

$$m\frac{d^3y}{dt^3} = \frac{He}{m}\left(Xe - He\frac{dy}{dt}\right).$$

The solutions of these equations, if x, y, dx/dt, dy/dt all vanish when $t=0$, is expressed by

$$y = \frac{Xm}{eH^2}\left\{\frac{e}{m}Ht - \sin\left(\frac{e}{m}Ht\right)\right\},$$

$$x = \frac{Xm}{eH^2}\left\{1 - \cos\left(\frac{e}{m}Ht\right)\right\}.$$

The equations show that the path of the particle is a cycloid, the generating circle of which has a diameter equal to $2Xm/eH^2$, and rolls on the line $x=0$.

Suppose now that we have a metal plate AB exposed to ultra-violet light, placed parallel to a larger metal plate CD perforated so as to allow the light to pass through it and fall upon the plate AB. Then, if CD is at a higher electric potential than AB, all the negatively electrified particles which start from AB will reach CD if this plate is large compared with AB, the particles travelling along the lines of electric force. Let us now suppose that a uniform magnetic force equal to H, and at right angles to the electric force, acts on the particles; these particles will now describe cycloids and will reach a distance $2Xm/eH^2$ from the place from which they start, and after reaching this distance they will again approach the plate. Thus if the plate CD is distant from AB by less than $2Xm/eH^2$, every particle which leaves AB will reach CD provided CD stretches forward enough to prevent the particles passing by on one side. Now the distance parallel to y through which the particle has travelled when it is at the greatest distance from AB is $\pi Xm/eH^2$: hence if CD stretches beyond AB by this distance at least, all the particles will be caught by CD and the magnetic field will produce no diminution in the rate of leak between AB and CD. If, on the other hand, the distance between the plates is greater than $2Xm/eH^2$, then a particle starting from AB will turn back before it reaches CD: it will thus never reach it, and the rate at which CD acquires negative electrification will be diminished by the magnetic force. Hence, if this view of the action of the magnetic field is correct, if we begin with the plates very near together and gradually increase the distance between them, we should expect that, at first with the plates quite close together, the rate at which CD received a negative charge would not be affected by the magnetic force, but as soon as the distance between the plates was equal to $2\mathrm{X}\,m\,e\mathrm{H}^2$ the magnetic force would greatly diminish the rate at which CD received a negative charge, and would in fact reduce the rate almost to zero if all the negatively electrified particles came from the surface of AB. Hence, if we measure the distance between the plates when the magnetic force first diminishes the rate at which CD receives a negative charge, we shall determine the value of $2\mathrm{X}m/e\mathrm{H}^2$; and as we can easily determine X and H, we can deduce the value of m/e.

The way in which this method was carried into practice was as follows, the apparatus being shown in fig. 1.

Fig. 1.

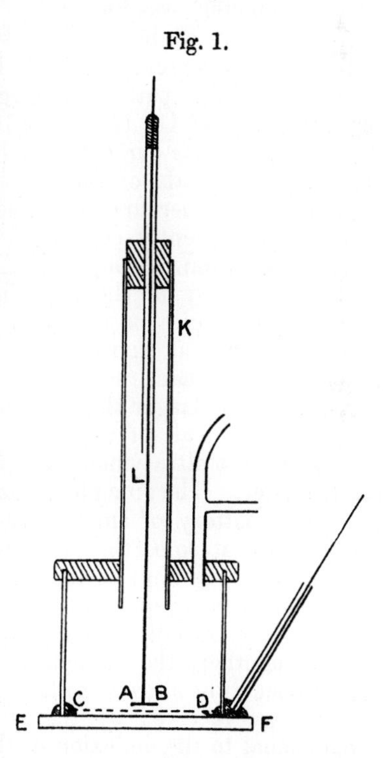

AB is a carefully polished zinc plate about 1 centim. in diameter, while CD is a grating composed of very fine wires crossing each other at right angles, the ends being soldered into a ring of metal; the wires formed a network with a mesh about 1 millim. square. This was placed parallel to AB on the quartz plate EF, which was about 4 millim. thick. The grating was very carefully insulated. The system was enclosed in a glass tube which was kept connected with a mercury-pump provided with a McLeod gauge. The ultra-violet light was supplied from an arc about 3 millim. long between zinc terminals. The induction-coil giving the arc was placed in a metal box, and the light passed through a window cut in the top of the box; over this window the quartz base of the vessel was placed, a piece of wire gauze connected with the earth being placed between the quartz and the window. The plate AB was carried by the handle L which passed through a sealing-wax stopper in the tube K. The magnet used was an electromagnet of the horseshoe type. The magnetic force due to the magnet was determined by observing the deflexion of a ballistic galvanometer when an exploring coil, of approximately the same vertical dimension as the distance between the plates AB and CD, was withdrawn from between its poles. The coil was carefully placed so as to occupy the same part of the magnetic field as that occupied by the space between AB and CD when the magnet was used to affect the rate of leak of electricity between AB and CD. In this way the intensity of the magnetic field between the poles of the magnet was determined for a series of values of the current through the magnetizing-coils of the electromagnet ranging between 1 and 4·5 amperes, and a curve was drawn which gave the magnetic force when the magnetizing-current (observed by an amperemeter) was known.

The pressure of the gas in the tube containing the plate was reduced by the mercury-pump to 1/100 of a millim. of mercury. As the mean free path of hydrogen molecules at atmospheric pressure and 0° C. is $1{\cdot}85 \times 10^{-5}$ centim. (Emil Meyer, *Kinetische Theorie der Gase*, p. 142), and of air 10^{-5} centim., the mean free paths of these gases at the pressure of 1/100 of a millim. of mercury are respectively 14 and 7·6 millim., and are consequently considerably greater than the greatest distance, 4 millim., through which the electrified particles have to travel in any of the experiments. These are the free paths for molecules of the gas ; if, as we shall see reason to believe, the actual carriers of the negative electrification are much smaller than the molecules, the free paths of these carriers will be larger than the numbers we have quoted.

The rate of leak of negative electricity to CD when AB was exposed to ultra-violet light was measured by a quadrant-electrometer. The zinc plate was connected with the negative pole of a battery of small storage-cells, the positive pole of which was put to earth. One pair of the quadrants of the electrometer was kept permanently connected with the earth, the other pair of quadrants was connected with the wire gauze CD. Initially the two pairs of quadrants were connected together, the connexion was then broken, and the ultra-violet light allowed to fall on the zinc plate ; the negative charge received by the wire gauze in a given time is proportional to the deflexion of the electrometer in that time. By this method the following results were obtained: when the difference of potential between the illuminated plate and the wire gauze was greater than a certain value, depending upon the intensity of the magnetic force and the distance between AB and CD, no diminution in the deflexion of the electrometer was produced by the magnetic field, in fact in some cases the deflexion was just a little greater in the magnetic field. The theory just given indicates that the deflexion

2 Q 2

ought to be the same: the small increase (amounting to not more than 3 or 4 per cent.) may be due to the obliquity of the path of the particles in the magnetic field, causing more of them to be caught by the wires of the grating than would be the case if the paths of the particles were at right angles to the plane of the gauze. When the difference of potential is reduced below a certain value, the deflexion of the electrometer is very much reduced by the magnetic field; it is not, however, at once entirely destroyed when the potential-difference passes through the critical value. The simple theory just given would indicate a very abrupt transition from the case when the magnetic force produces no effect, to that in which it entirely stops the flow of negative electricity to CD. In practice, however, I find that the transition is not abrupt: after passing a certain difference of potential the diminution in the electric charge received by CD increases gradually as the potential-difference is reduced, and there is not an abrupt transition from zero effect to a complete stoppage of the leak between AB and CD. I think this is due to the ionization not being confined to the gas in contact with the illuminated plate, but extending through a layer of gas whose thickness at very low pressures is quite appreciable. The existence of a layer of this kind is indicated by an experiment of Stoletow's. Stoletow found that the maximum current between two plates depended at low pressures to a considerable extent upon the distance between the plates, increasing as the distance between the plates was increased. Now the maximum current is the one that in one second uses up as many ions as are produced in that time by the ultra-violet light. If all the ions are produced close to the illuminated plate, increasing the distance between the plates will not increase the number of ions available for carrying the current; if, however, the ions are produced in a layer of sensible thickness, then, until the distance between the plates exceeds the thickness of this layer, an increase in the distance between the plates will increase the number of ions, and so increase the maximum current. If this layer has a sensible thickness, then the distance d which has to be traversed by the ions before reaching the gauze connected with the electrometer ranges from the distance between the plates to the difference between this distance and the thickness of the layer. The first ions to be stopped by the magnetic field will be those coming from the surface of the illuminated plate, as for these d has the greatest value: hence we may use the equation

$$d=\frac{2Xm}{eH^2}, \quad . \quad . \quad . \quad . \quad . \quad . \quad . \quad . \quad (1)$$

if d represents the distance between the plates, X the value of the electric field when the rate of leak first begins to be affected by the magnetic force H. Assuming that the field is uniform,

$$X = V/d,$$

where V is the potential-difference between the plates ; and equation (1) becomes

$$d^2 = \frac{2Vm}{eH^2}.$$

The negative ions travelling between the plates will disturb to some extent the uniformity of the field between the plates ; but if the intensity of the ultra-violet light is not too great, so that the rate of leak and the number of ions between the plates is not large, this want of uniformity will not be important. A calculation of the amount of variation due to this cause showed that its effect was not large enough to make it worth while correcting the observations for this effect, as the variation in the intensity of the ultra-violet light was sufficient to make the errors of experiments much larger than the correction.

The following is a specimen of the observations :—

Distance between the plates ·29 centim.

Strength of magnetic field 164. Pressure 1/100 millim.

Potential-difference between Plates, in volts.	Deflexion of Electrometer in 30 secs.	
	Magnet off.	Magnet on.
240	180	190
120	160	165
80	160	140
40	130	75

These observations showed that the critical value of the potential-difference was about 80 volts. A series of observations were then made with potential-differences increasing from 80 volts by 2 volts at a time, and it was found that 90 volts was the largest potential-difference at which any effect due to the magnet could be detected. The results of a number of experiments are given in the following table :—

d (in cm.).	H.	V in absolute measure.	e/m.
·18	170	40×10^8	$8{\cdot}5\times10^6$
·19	170	30×10^8	$5{\cdot}8\times10^6$
·20	181	46×10^8	$7{\cdot}0\times10^6$
·29	167	84×10^8	$7{\cdot}1\times10^6$
·29	164	90×10^8	$7{\cdot}6\times10^6$
·30	160	86×10^8	$7{\cdot}4\times10^6$
·45	100	80×10^8	$7{\cdot}9\times10^6$

giving a mean value for e/m equal to $7{\cdot}3\times10^6$. The value I found for e/m for the cathode rays was 5×10^6; the value found by Lenard was $6{\cdot}4\times10^6$. Thus the value of e/m in the case of the convection of electricity under the influence of ultra-violet light is of the same order as in the case of the cathode rays, and is very different from the value of e/m in the case of the hydrogen ions in ordinary electrolysis when it is equal to 10^4. As the measurements of e, the charge carried by the ions produced by ultra-violet light to be described below, show that it is the same as e for the hydrogen ion in electrolyis, it follows that the mass of the carrier in the case of the convection of negative electricity under the influence of ultra-violet light is only of the order of 1/1000 of that of the hydrogen atom. Thus with ultra-violet light, as with cathode rays, the negative electrification at low pressures is found associated with masses which are exceedingly small fractions of the smallest mass hitherto known—that of the hydrogen atom.

I have examined another case in which we have convection of electricity at low pressures by means of negatively electrified particles—that of the discharge of electricity produced by an incandescent carbon filament in an atmosphere of hydrogen. In this case, as Elster and Geitel (Wied. *Ann.* xxxviii. p. 27) have shown, we have negative ions produced in the neighbourhood of the filament, and the charge on a positively electrified body in the neighbourhood of the filament is discharged by these ions, while if the body is negatively electrified it is not discharged. If the filament is negatively, and a neighbouring body positively electrified, there will be a current of electricity between the filament and the body, while there will be no leak if the filament is positively and the body negatively electrified. Elster and Geitel (Wied.

Ann. xxxviii. p. 27) showed that the rate of leak from a negatively electrified filament was at low pressures diminished by the action of the magnetic field. On the theory of charged ions, the effect of the magnet in diminishing the rate of leak could be explained in the same way as the effect on the convection due to ultra-violet light. A series of experiments were made which showed that the effects due to the magnetic field were consistent with this explanation, and led to a determination of e/m for the carriers of the negative electricity.

The apparatus was of the same type as that used in the preceding experiments. The wire gauze and the zinc plate were replaced by two parallel aluminium disks about 1·75 centim. in diameter; between these disks, and quite close to the upper disk, there was a small semicircular carbon filament which was raised to a red heat by the current from four storage-cells. The carbon filament was placed close to the axis of the disks; the object of the upper disk was to make the electric field between the disks more uniform. The lower plate was connected with the electrometer. The plates and filaments were enclosed in a glass tube which was connected with a mercury-pump, by means of which the pressure, after the vessel had been repeatedly filled with hydrogen, was reduced to ·01 millim. of mercury. Great difficulty was found at first in getting any consistent results with the incandescent carbon filament: sometimes the filament would discharge positive as well as negative electricity; indeed sometimes it would discharge positive and not negative. Most of these irregularities were traced to gas given out by the incandescent filament; and it was found that by keeping the filament almost white-hot for several hours, and continually pumping and refilling with hydrogen, and then using the filament at a much lower temperature than that to which it had been raised in this preliminary heating, the irregularities were nearly eliminated, and nothing but negative electrification was discharged from the filament. When this state was attained, the effect of magnetic force showed the same characteristics as in the case of ultra-violet light. When the difference of potential between the filament and the lower plate was small, the effect of the magnetic force was very great, so much so as almost to destroy the leak entirely; when, however, the potential-difference exceeded a certain value, the magnetic force produced little or no effect upon the leak. An example of this is shown by the results of the following experiment:—

The distance between the carbon filament and the plate connected with the electrometer was 3·5 millim., the strength of the magnetic field 170 C.G.S. units.

Difference of Potential between wire and plate, in volts.	Leak in 5 seconds.		Ratio of leaks.
	Without magnetic field.	With magnetic field.	
40	43	1	·023
80	170	50	·29
120	300	250	·83
140	345	345	1·0
160	400	430	1·07

Taking 140 volts as the critical value of the potential-difference, we find by equation (1) that

$$\frac{e}{m} = 7\cdot 8 \times 10^6.$$

The results of this and similar experiments are given in the following table; V denoting the critical potential-difference in C.G.S. units, and H the magnetic force:

d.	V.	H.	*e/m.*
·35	140×10^8	170	$7\cdot8 \times 10^6$
·35	220×10^8	220	$7\cdot5 \times 10^6$
·35	170×10^8	170	$9\cdot6 \times 10^6$
·35	130×10^8	170	$7\cdot2 \times 10^6$
·35	120×10^8	120	$11\cdot3 \times 10^6$

giving $8\cdot 7 \times 10^6$ as the mean value of *e/m*. This value does not differ much from that found in the case of ultra-violet light. In the case of the incandescent filament the ions are only produced at a small part of the plate, and not over the whole surface as in the case of ultra-violet light, so the conditions do not approximate so closely to those assumed in the theory. We conclude that the particles which carry the negative electrification in this case are of the same nature as those which carry it in the cathode rays and in the electrification arising from the action of ultra-violet light.

The unipolar positive leak which occurs from an incandescent platinum wire in air or oxygen, and in which the moving bodies are positively electrified, was found not to be affected by a magnetic field of the order of that used in the experiments on the negative leak. This had already been observed by Elster and Geitel (Wied. *Ann.* xxxviii. p. 27).

On the theory of the effect given in this paper, the absence of magnetic effect on the positively charged carriers indicates that e/m is much smaller or m/e much larger for the positive ions than it is for the negative. I am engaged with some experiments on the effect of the magnetic field on the convection of electricity by positive ions, using very strong magnetic fields produced by a powerful electromagnet kindly lent to me by Professor Ewing. From the results I have already got, it is clear that m/e for the positive ions produced by an incandescent wire must be at least 1000 times the value for the negative ions, and this is only an inferior limit.

The positive and negative ions produced by incandescent solids show the same disproportion of mass as is shown by the positive and negative ions in a vacuum-tube at low pressures.

W. Wien (Wied. *Ann.* lxv. p. 440) and Ewers (Wied. *Ann.* lxix. p. 187) have measured the ratio of m/e for the positive ions in such a tube, and found that it is of the same order as the value of m/e in ordinary electrolysis; Ewers has shown that it depends on the metal of which the cathode is made. Thus the carriers of positive electricity at low pressures seem to be ordinary molecules, while the carriers of negative electricity are very much smaller.

Measurement of the Charge on the Ion produced by the Action of Utra-Violet Light on a Zinc Plate.

This charge was determined by the method used by me to measure the charge on the ions produced by the action of Röntgen rays on a gas (Phil. Mag. Dec. 1898); for the details of the method I shall refer to my former paper, and here give only an outline of the principle on which the method is based. Mr. C. T. R. Wilson (Phil. Trans. 1898) discovered that the ions produced by ultra-violet light act like those produced by Röntgen rays, in forming nuclei around which water will condense from dust-free air when the supersaturation exceeds a certain definite value.

Suppose, then, we wish to find the number of ions produced by ultra-violet light in a cubic centimetre of air. We cool the air by a sudden expansion until the supersaturation produced by the cooling is sufficient to form a cloud round the ions: the problem of finding the number of ions per cub. centim. is now reduced to finding the number of drops per cub. centim. in this cloud. We can do this in the following way:—If we know the amount of the expansion we can calculate the amount of water deposited per cub. centim. of the cloud; this water is deposited as drops, and if the drops are of equal size, the number of drops per cub. centim. will be equal to the volume of water per cub. centim. divided by the volume of one of the drops. Hence, if we know the size of the drops, we can calculate the number. The size of the drops in the cloud was determined by observing v, the velocity with which they fall under gravity, and then deducing a, the radius of the drop, by means of the equation

$$v = \frac{2}{9}\frac{ga^2}{\mu},$$

where μ is the coefficient of viscosity of the gas through which the drop falls.

In this way we can determine n the number of ions per cub. centim.: if e is the charge on an ion, v the velocity with which it moves under a known electric force, the quantity of electricity which crosses unit area in unit time under this force is equal to neu. We can determine this quantity if we allow the negative ions to fall on a plate connected with a condenser of known capacity and measure the rate at which the potential falls. We thus determine the product neu, and we already know n; u has been determined by Mr. Rutherford (Proc. Camb. Phil. Soc. ix. p. 401); for air at atmospheric pressure u is proportional to the potential gradient, and when this is one volt per centim., u is 1·5 centim. per second; for hydrogen at atmospheric pressure u is 4·5 centim. per second for the same potential gradient. Hence, as in the known product neu we know n and u, we can deduce the value of e the charge on the ion.

There are some features in the condensation of clouds by ultra-violet light which are not present in the clouds formed by the Röntgen rays. In the first place, the cloud due to the ultra-violet light is only formed in an electric field. When there is no electric field, the ions remain close to the surface of the illuminated plate, and are not diffused through the region in which the cloud has to be formed; to get the negative ions into this region we must electrify the plate negatively; when this is done, expansion produces a cloud. Again, if the ultra-violet light is very strong, Mr. C. T. R. Wilson has shown (Phil. Trans. 1899) that large nuclei are produced in the gas through which the light passes; these are distinct from those produced near a metal plate on which the light falls, and they can produce a cloud with very little supersaturation; these nuclei are not ions, for they do not move in an electric field, and the drops formed round these nuclei ought therefore not to be counted in estimating the number of negative ions. For this reason it is necessary to use ultra-violet light

Fig. 2.

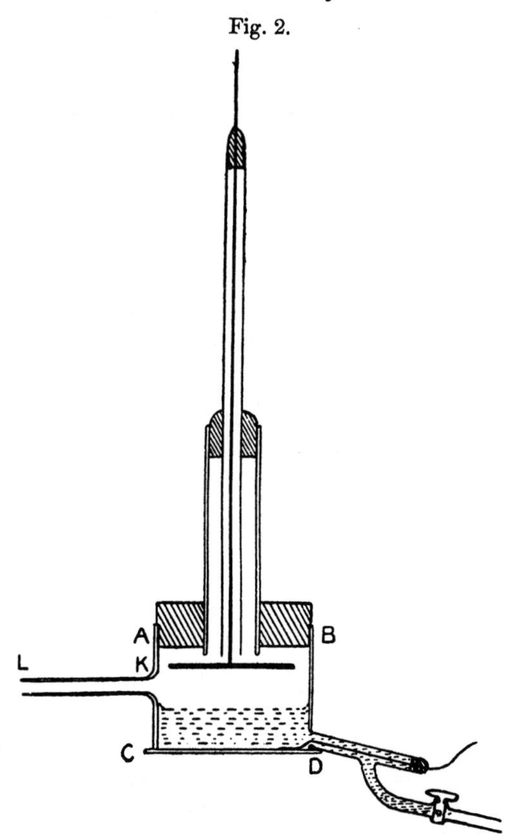

of small intensity, and there are in addition other reasons which make it impossible to work with strong light. I found when working with the ions produced by Röntgen rays, that it was impossible to get good results unless the rays were weak and the clouds therefore thin. If the rays were strong, one expansion was not sufficient to bring down all the ions by the cloud; sometimes as many as five or six expansions were required to remove the ions from the vessel. Another reason why the strong rays do not give good results is that there are slight convection-currents in the vessel after the expansion, for the walls of the vessel are warmer than the gas; this gives rise to convection-currents in the gas, the gas going up the sides and down the middle of the vessel. The velocity of the convection-current is added on to the velocity of the ions due to gravity; and if the velocity of the ions is very small, as it is when the rays are strong and the drops numerous, a very small convection-current will be sufficient to make the actual rate of fall of the drops very different from that of a drop of the same size falling through air at rest. All the reasons are operative in the case of ultra-violet light, and it is only when the intensity of the light is small that I have got consistent results.

The vessel in which the expansion took place is shown in fig. 2. AB is a glass tube about 3·6 cm. in diameter; the base CD is a quartz plate about ·5 cm. thick; on the top of this there is a layer of water in electrical connexion with the earth about 1 cm. in thickness; the illuminated zinc plate was 3·2 cm. in diameter, and was 1·2 cm. above the surface of the water. The ultra-violet light was produced by an arc about ·3 cm. long, between zinc terminals connected with an induction-coil; the arc was about 40 cm. below the lower face of the quartz plate. The space between the zinc plate and the water surface was illuminated by an arc-light so as to allow the rate of fall of the drops to be accurately measured. The tube LK connected this vessel with the apparatus used in the previous experiments; a figure of this is given in the Phil. Mag. Dec. 1898.

To observe the current of electricity through the gas, the illuminated plate was connected with one pair of quadrants of an electrometer, the other pair of quadrants being kept connected with the earth. The capacity C of the system, consisting of the plate, connecting wires and quadrants of the electrometer, was determined. The plate was then charged to a negative potential, and the deflexion of the electrometer-needles observed. The induction-coil was now set in action, and the ultra-violet light allowed to fall on the zinc plate: the deflexion of the electrometer-needle immediately began to decrease; the rate at which it decreased was determined by measuring the diminution of the deflexion in 30 seconds.

Let D be the original deflexion of the electrometer, let this correspond to a potential-difference equal to αD between the plate and the earth. If b is the distance between the zinc plate and the surface of the water, the potential gradient is $\alpha D/b$. If A is the area of the plate, n the number of ions per cub. centim., e the charge on an ion, u_0 the velocity of the ion under unit potential gradient, then the quantity of negative electricity lost by the plate in one second is

$$\mathrm{A}neu_0\alpha \mathrm{D}/b.$$

But the plate is observed to fall in potential by αb per second, and the capacity of the system attached to the plate is C: hence the loss of electricity by the plate per second is

$$\mathrm{C}\alpha d.$$

Equating these two expressions for the loss of electricity, we get

$$\mathrm{A}neu_0\alpha \mathrm{D}/b = \mathrm{C}\alpha d$$

or

$$e = \frac{b}{nu_0}\frac{\mathrm{C}}{\mathrm{A}}\frac{d}{\mathrm{D}}.$$

Hence knowing b, C, A, and u_0, if we measure n and d/D we can determine e.

To calculate n we begin by finding the volume of water deposited in consequence of the expansion in each cub. centim. of the expansion. In my previous paper I show how this can be determined if we know the ratio of the final to the initial volumes and the temperature before expansion. In the present experiments the final volume was 1·36 times the initial volume, and the temperature before expansion was 18°·5 C. It follows from this that 50×10^{-7} cub. centim. of water were deposited in each cub. centim. of the expansion chamber.

If a is the radius of one of the drops, the volume of a drop is $4\pi a^3/3$, and hence $n' = \dfrac{3 \times 50 \times 10^{-7}}{4\pi a^2}$: here n' is the number of ions per cub. centim. of the expanded gas.

If v is the velocity of fall

$$v = \frac{2}{9}\frac{ga^2}{\mu}.$$

Since for air $\mu = 1{\cdot}8 \times 10^{-4}$, we find

$$a = \frac{v^{\frac{1}{2}}}{1{\cdot}1 \times 10^3},$$

and

$$\frac{4}{3}\pi a^3 = 3{\cdot}14\, v^{\frac{3}{2}} \times 10^{-9},$$

$$n' = \frac{5000}{3{\cdot}14\, v^{\frac{3}{2}}}.$$

This is the number in 1 cub. centim. of the expanded gas; the number in 1 cub. centim. of the gas before expansion is $1{\cdot}36\,n'$. To find n the number of ions we must subtract from $1{\cdot}36\,n'$ the number of drops which are formed when the ultra-violet light does not fall on the plate. With an expansion as large as 1·36, Mr. Wilson has shown that a few drops are always formed in dust-free air, even when free from the influence of Röntgen rays or ultra-violet light. If V be the velocity with which these drops formed in the absence of the light fall, then the number of drops due to these nuclei is

$$\frac{1{\cdot}36 \times 5000}{3{\cdot}14\,V^{\frac{3}{2}}}.$$

Subtracting this from $1{\cdot}36\,n'$, we find

$$n = 2{\cdot}07 \times 10^3 \left\{ \frac{1}{v^{\frac{3}{2}}} - \frac{1}{V^{\frac{3}{2}}} \right\}.$$

In making this correction we have assumed that the clouds form round these nuclei even when the negative ions due to the ultra-violet light are present. If the cloud formed more readily about the negative ions than about the nuclei, the ions would rob the nuclei of their water, and we should not need the correction. The following table gives the result of some experiments; in making the observation on the cloud the same potential-difference between the plate and the water was used as when observing the value of d/D: u_0 was determined by Prof. Rutherford as $1{\cdot}5 \times 3 \times 10^2$, and A was $\pi(1{\cdot}6)^2$ throughout the experiments.

b.	C.	d/D.	v.	V.	$e \times 10^{10}$.
1·2	62	·0017	·13	·3	7·9
1·2	62	·0019	·11	·3	7·3
·9	50	·0012	·14	·3	5·3
1·2	65	·0035	·08	·3	7·3
1·2	50	·0018	·11	·3	6
1·2	40	·0018	·14	·3	7

The mean value of e is $6{\cdot}8 \times 10^{-10}$. The values differ a good deal, but we could not expect a very close agreement unless we could procure an absolutely constant source of ultra-violet light, as these experiments are very dependent on the constancy of the light; since the electrical part of the experiment measures the average intensity of the light over 30 seconds, while the observations on the cloud measure the intensity over an interval of a small fraction of a second.

The value of e found by me previously for the ions produced by Röntgen rays was $6{\cdot}5\times10^{-8}$: hence we conclude that e for the ions produced by ultra-violet light is the same as e for the ions produced by the Röntgen rays ; and as Mr. Townsend has shown that the charge on these latter ions is the same as the charge on an atom of hydrogen in electrolysis, we arrive at the result previously referred to, that the charge on the ion produced by ultra-violet light is the same as that on the hydrogen ion in ordinary electrolysis.

The experiments just described, taken in conjunction with previous ones on the value of m/e for the cathode rays (J. J. Thomson, Phil. Mag. Oct. 1897), show that in gases at low pressures negative electrification, though it may be produced by very different means, is made up of units each having a charge of electricity of a definite size ; the magnitude of this negative charge is about 6×10^{-10} electrostatic units, and is equal to the positive charge carried by the hydrogen atom in the electrolysis of solutions.

In gases at low pressures these units of negative electric charge are always associated with carriers of a definite mass. This mass is exceedingly small, being only about $1{\cdot}4\times10^{-3}$ of that of the hydrogen ion, the smallest mass hitherto recognized as capable of a separate existence. The production of negative electrification thus involves the splitting up of an atom, as from a collection of atoms something is detached whose mass is less than that of a single atom. We have not yet data for determining whether the mass of the negative atom is entirely due to its charge. If the charge is e, the apparent mass due to the charge supposed to be collected on a sphere of radius a is $\frac{1}{3}e^2/\mu a$: hence m/e in this case is $e/3\mu a$. Substituting the values of m/e and e found above, we find that a would be of the order 10^{-13} centim.

We have no means yet of knowing whether or not the mass of the negative ion is of electrical origin. We could probably get light on this point by comparing the heat produced by the bombardment by these negatively electrified particles of the inside of a vessel composed of a substance transparent to Röntgen rays, with the heat produced when the vessel was opaque to those rays. If the mass was "mechanical," and not electrical, the heat produced should be same in the two cases. If, on the other hand, the mass were electrical, the heat would be less in the first case than in the second, as part of the energy would escape through the walls.

Hitherto we have been considering only negative electrification; as far as our present knowledge extends positive electrification is never associated with masses as small as those which invariably accompany negative electrification in gases at low pressures. From W. Wien's experiments on the ratio of the mass to the electric charge for the carriers of positive electrification in a highly exhausted vacuum-tube (Wied. *Ann.* lxv. p. 440), it would seem that the masses with which positive electrification is associated are comparable with the masses of ordinary atoms. This is also in accordance with the experiments of Elster and Geitel (Wied. *Ann.* xxxviii. p. 27), which show that when positive ions are produced by an incandescent platinum wire in air they are not affected to anything like the same extent as negative ions produced by an incandescent carbon filament in hydrogen.

It is necessary to point out that the preceding statements as to the masses of the ions are only true when the pressure of the gas is very small, so small that we are able to determine the mass of the carriers before they have made many collisions with the surrounding molecules. When the pressure is too high for this to be the case, the electric charge, whether positive or negative, seems to act as a nucleus around which several molecules collect, just as dust collects round an electrified body, so that we get an aggregate formed whose mass is larger than that of a molecule of a gas.

The experiments on the velocities of the ions produced by Röntgen or uranium rays, by ultra-violet light, in flames or in the arc, show that in gases at pressures comparable with the atmospheric pressure, the electric charges are associated with masses which are probably several times the mass of a molecule of the gas, and enormously greater than the mass of a carrier of negative electrification in a gas at a low pressure.

There are some other phenomena which seem to have a very direct bearing on the nature of the process of ionizing a gas. Thus I have shown (Phil. Mag. Dec. 1898) that when a gas is ionized by Röntgen rays, the charges on the ions are the same whatever the nature of the gas: thus we get the same charges on the ions whether we ionize hydrogen or oxygen. This result has been confirmed by J. S. Townsend ("On the Diffusion of Ions," Phil. Trans. 1899), who used an entirely different method. Again, the ionization of a gas by Röntgen rays is in general an additive property; *i. e.*, the ionization of a compound gas AB, where A and B represent the atoms of two elementary gases, is one half the sum of the ionization of A_2 and B_2 by rays of the same intensity, where

A_2 and B_2 represent diatomic molecules of these gases (Proc. Camb. Phil. Soc. vol. x. p. 9). This result makes it probable that the ionization of a gas in these cases results from the splitting up of the atoms of the gas, rather than from a separation of one atom from the other in a molecule of the gas.

These results, taken in conjunction with the measurements of the mass of the negative ion, suggest that the ionization of a gas consists in the detachment from the atom of a negative ion; this negative ion being the same for all gases, while the mass of the ion is only a small fraction of the mass of an atom of hydrogen.

From what we have seen, this negative ion must be a quantity of fundamental importance in any theory of electrical action; indeed, it seems not improbable that it is the fundamental quantity in terms of which all electrical processes can be expressed. For, as we have seen, its mass and its charge are invariable, independent both of the processes by which the electrification is produced and of the gas from which the ions are set free. It thus possesses the characteristics of being a fundamental conception in electricity; and it seems desirable to adopt some view of electrical action which brings this conception into prominence. These considerations have led me to take as a working hypothesis the following method of regarding the electrification of a gas, or indeed of matter in any state.

I regard the atom as containing a large number of smaller bodies which I will call corpuscles; these corpuscles are equal to each other; the mass of a corpuscle is the mass of the negative ion in a gas at low pressure, *i. e.* about 3×10^{-26} of a gramme. In the normal atom, this assemblage of corpuscles forms a system which is electrically neutral. Though the individual corpuscles behave like negative ions, yet when they are assembled in a neutral atom the negative effect is balanced by something which causes the space through which the corpuscles are spread to act as if it had a charge of positive electricity equal in amount to the sum of the negative charges on the corpuscles. Electrification of a gas I regard as due to the splitting up of some of the atoms of the gas, resulting in the detachment of a corpuscle from some of the atoms. The detached corpuscles behave like negative ions, each carrying a constant negative charge, which we shall call for brevity the unit charge; while the part of the atom left behind behaves like a positive ion with the unit positive charge and a mass large compared with that of the negative ion. On this view, electrification essentially involves the splitting up of the atom, a part of the mass of the atom getting free and becoming detached from the original atom.

A positively electrified atom is an atom which has lost some of its "free mass," and this free mass is to be found along with the corresponding negative charge. Changes in the electrical charge on an atom are due to corpuscles moving from the atom when the positive charge is increased, or to corpuscles moving up to it when the negative charge is increased. Thus when anions and cations are liberated against the electrodes in the electrolysis of solutions, the ion with the positive charge is neutralized by a corpuscle moving from the electrode to the ion, while the ion with the negative charge is neutralized by a corpuscle passing from the ion to the electrode. The corpuscles are the vehicles by which electricity is carried from one atom to another.

We are thus led to the conclusion that the mass of an atom is not invariable: that, for example, if in the molecule of HCl the hydrogen atom has the positive and the chlorine atom the negative charge, then the mass of the hydrogen atom is less than half the mass of the hydrogen molecule H_2; while, on the other hand, the mass of the chlorine atom in the molecule of HCl is greater than half the mass of the chlorine molecule Cl_2.

The amount by which the mass of an atom may vary is proportional to the charge of electricity it can receive; and as we have no evidence that an atom can receive a greater charge than that of its ion in the electrolysis of solutions, and as this charge is equal to the valency of the ion multiplied by the charge on the hydrogen atom, we conclude that the variability of the mass of an atom which can be produced by known processes is proportional to the valency of the atom, and our determination of the mass of the corpuscle shows that this variability is only a small fraction of the mass of the original atom.

In the case of the ionization of a gas by Röntgen or uranium rays, the evidence seems to be in favour of the view that not more than one corpuscle can be detached from any one atom. For if more than one were detached, the remaining part of the atom would have a positive charge greater than the negative charge carried by each of the detached corpuscles. Now the ions, in virtue of their charges, act as nuclei around which drops of water condense when moist dust-free gas is suddenly expanded. If the positive charge were greater than the individual negative ones, the positive ions would be more efficient in producing cloudy condensation than the negative one, and would give a cloud with smaller expansion. As a matter of fact, however, the reverse is the case, as C. T. R. Wilson (Phil. Trans. 1899) has shown that it requires a considerably greater expansion to produce a cloud in dust-free air on positive ions than on negative ones when the ions are produced by Röntgen rays.

Though only a small fraction of the mass of an atom can be detached by any known process, it does not follow that the part left behind does not contain more corpuscles which could be detached by more powerful means than we have hitherto been able to use. For it is evident that it will require a greater expenditure of energy to tear two corpuscles from one atom than to tear two corpuscles one from each of two separate atoms; for when one corpuscle has been torn off from an atom the atom is positively electrified, and it will be more difficult to tear off a second negatively electrified corpuscle from this positively electrified atom, than it was to tear the first from the originally neutral atom. A reason for believing that there are many more corpuscles in the atom than the one or two that can be torn off, is afforded by the Zeeman effect. The ratio of the mass to the charge, as determined by this effect, is of the same order as that we have deduced from our measurements on the free corpuscles; and the charges carried by the moving particles, by which the Zeeman effect is explained, are all negatively electrified. Now, if there were only one or two of these corpuscles in the atom, we should expect that only one or two lines in the spectrum would show the Zeeman effect; for even if the coordinates fixing the position of the moving corpuscles were not "principal coordinates," though there might be a secondary effect on the periods of the other oscillations due to their connexion with these coordinates, yet we should expect this secondary effect to be of quite a different order from the primary one. As, however, there are a considerable number of lines in the spectrum which show Zeeman effects comparable in intensity, we conclude that there are a considerable number of corpuscles in the atom of the substance giving this spectrum.

I have much pleasure in thanking my assistant Mr. E. Everett for the help he has given me in making the experiments described in this paper.

6

Later Years

The Electromagnetic View of Matter and the Thomson Atom

Around 1900, Thomson felt that his corpuscle idea was beginning to be accepted, following his 1899 paper on the charge on corpuscles from the photo-electric effect and his talk at the British Association meeting in Dover. His theory of discharge by dissociation was, by now, fairly well elucidated and was rapidly and widely accepted. He published his book *Conduction of Electricity Through Gases* in 1903, and was awarded the Nobel Prize in 1906 for his work on discharge. For the annual dinner that year, A.A. Robb wrote a new song:

AN EMANATION

Air: 'Chon-Kina' (Geisha)

1. I'm the smartest of professors in the town,
You should see me in my college cap and gown,
At some learned association
When I give the great oration
And I'm covered with a halo of renown.
O'er the scientific world I hold my sway,
And it harkens unto every word I say,
But there comes a dinner yearly,
T'is a thing I love most dearly
For they treat me there in quite another way.

Chorus:
John Joseph, John Joseph, John, John, Joseph Joseph.
That is was they call me at the dinner once a year.

2. The structure of the atoms I explain,
By the number of corpuscles they contain,
Which according to my notion
Are in very rapid motion,
Though the speed of light they never quite attain.
They've a mass two-thirds of e^2 over a,
But I'll give you more details another day,
For I'm feeling so elated
With the way I have been feted
That my sole desire is just to hear you say:
John, Joseph, etc.

3. For reasons which I'll leave you to surmise
'Twas decided I should get the Nobel prize,
So they called me off to Sweden,
And I thought I was in Eden,
O, you should have seen the sparkle in my eyes.
My diploma neatly fastened with a string
Was presented by His Majesty the King,
But no sooner was it over
Than I crossed the Straits of Dover
To be present at the dinner where they sing:
John, Joseph, etc.

As indicated in this song, Thomson now progressed to trying to bring new unity to physics based on corpuscle theory. As long ago as 1881 he had shown that the mass of a charged particle increased the faster it moved. Thomson explained this qualitatively by analogy with a sphere moving through water:

> When the sphere moves it sets the water around it in motion. The necessity of doing this makes the sphere behave as if its mass were increased by a mass equal to half the mass of a sphere of water of the same volume as the sphere itself. This additional mass is not in the sphere but in the space around it. . . If we adopt the electrical theory of the constitution of matter we may suppose that all mass is electrical in its origin, and therefore not in the atoms or molecules themselves but in the space around their charges. The hydrodynamical analogue of this would be the case of the motion through water of exceedingly thin spherical shells, so thin that their mass was infinitesimal in comparison with that of the water they displaced. A collection of these would have a finite mass equal to half that of the water they displaced. [1]

Now suppose that the whole of the mass of a corpuscle was due to its motion. Then there was the possibility of explaining all matter as electric charges (specifically corpuscles) in motion. The potential for unification was enormous and was seized on by many physicists, particularly in Britain and Germany.

Characteristically, Thomson himself did not pursue this topic in depth, preferring instead to try and bring as many phenomena as possible within the orbit of corpuscle theory. The recently discovered phenomena of X-rays and radioactivity presented obvious challenges within his own area of physics. As we have seen, the evidence radioactivity provided that atoms could split up and change their nature gave a strong impetus to acceptance of Thomson's corpuscle theory. But it did more: it encouraged Thomson to revive and pursue the atomic theory he had first suggested in 1897. How could the splitting up take place?

Thomson's theory has often been described as the 'plum pudding model' of the atom and set up as a straw man to be shot down in 1911 by Rutherford's evidence for a nuclear atom, the forerunner of our modern atomic model. Yet in its heyday, seven years before Rutherford suggested the existence of atomic nuclei, Thomson's was a sophisticated model which promised unification of physics based on the electromagnetic mass of the corpuscles, and it gave a natural explanation for valency, the periodic properties of the elements, and radioactivity.

The Thomson atom consisted of a sphere of positive electricity containing spinning rings of hundreds of corpuscles, the orientation and arrangement of which explained valency and whose eventual, but long delayed, instability gave a mechanism for radioactive decay. This atom model is described in *The Structure of the Atom*, reproduced on p. 215.

Thomson described his model and its possibilities in a letter to Rutherford of 18 February 1904:

> I have been working hard for some time at the structure of the atom, regarding the atom as built up of a number of corpuscles in equilibrium or steady motion under their mutual repulsions and a central attraction: it is surprising what a lot of interesting results come out. I really have hopes of being able to work out a reasonable theory of chemical combination and many other chemical phenomena. There are some interesting cases in which the stability of the atom depends on the rotation of the corpuscles, just as the stability of a spinning top depends on its rotation, the equilibrium ceasing to be stable when the velocity of the corpuscles round their orbit falls below a certain value, now a moving corpuscle is radiating energy so that its velocity is continually falling though very slowly, a time must come although it will be long in coming when the velocity falls below the critical value the previous arrangement becomes unstable, there is an explosion and some of the parts have their kinetic energy very much increased in fact they would behave somewhat like radium. I think a spinning top is a good illustration of the radium atom. [2]

The mass of this atom was entirely due to the corpuscles (hence the necessity for thousands of them), and Thomson always hoped to be able, eventually, to abandon the positive electrification, writing to Lodge on 11 April 1904:

> With regard to positive electrification I have been in the habit of using the crude analogy of a liquid with a certain amount of cohesion, enough to keep it from flying to bits under its own repulsion. I have however always tried to keep the physical conception of the positive electricity in the background because I have

> always had hopes (not yet realised) of being able to do without positive electrification as a separate entity and to replace it by some property of the corpuscles.
>
> When one considers that, all the positive electricity does, on the corpuscular theory, is to provide an attractive force to keep the corpuscles together, while all the observable properties of the atom are determined by the corpuscles one feels, I think, that the positive electrification will ultimately prove superfluous and it will be possible to get the effects we now attribute to it from some property of the corpuscle.
>
> At present I am not able to do this and I use the analogy of the liquid as a way of picturing the missing forces which is easily conceived and lends itself readily to analysis. [3]

As in 1897, and much earlier in 1882 with the vortex atom, Thomson drew an analogy to Mayer's magnets to suggest how the corpuscles might be arranged so as to explain the periodic properties of the elements. Some of his arrangements are shown in Figure 6.1.

Such systems of corpuscles would be unstable unless they were spinning rapidly; so Thomson set them spinning, thus making the analogy to Mayer's magnets even more difficult to draw exactly. Probably he envisaged the rings of spinning corpuscles much as he had the vortex rings of his early atomic theories. Radioactive decay now became a natural, even inevitable, consequence of his model: rotating corpuscles emit energy and slow down; the rate of

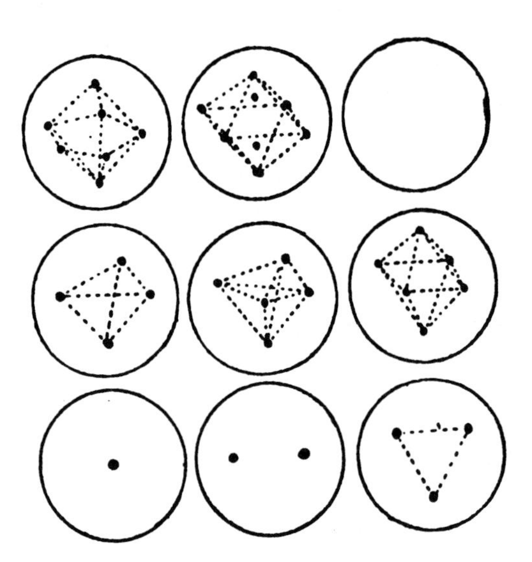

Figure 6.1 Thomson's suggested arrangements for a static system of corpuscles in an atom. Three-dimensional arrangements for one to eight corpuscles. He had worked these out mathematically. But his actual atomic model contained thousands of corpuscles orbiting rapidly, and for such a system analysis was extremely difficult. Instead, Thomson used an analogy to Mayer's magnets to suggest a periodic arrangement

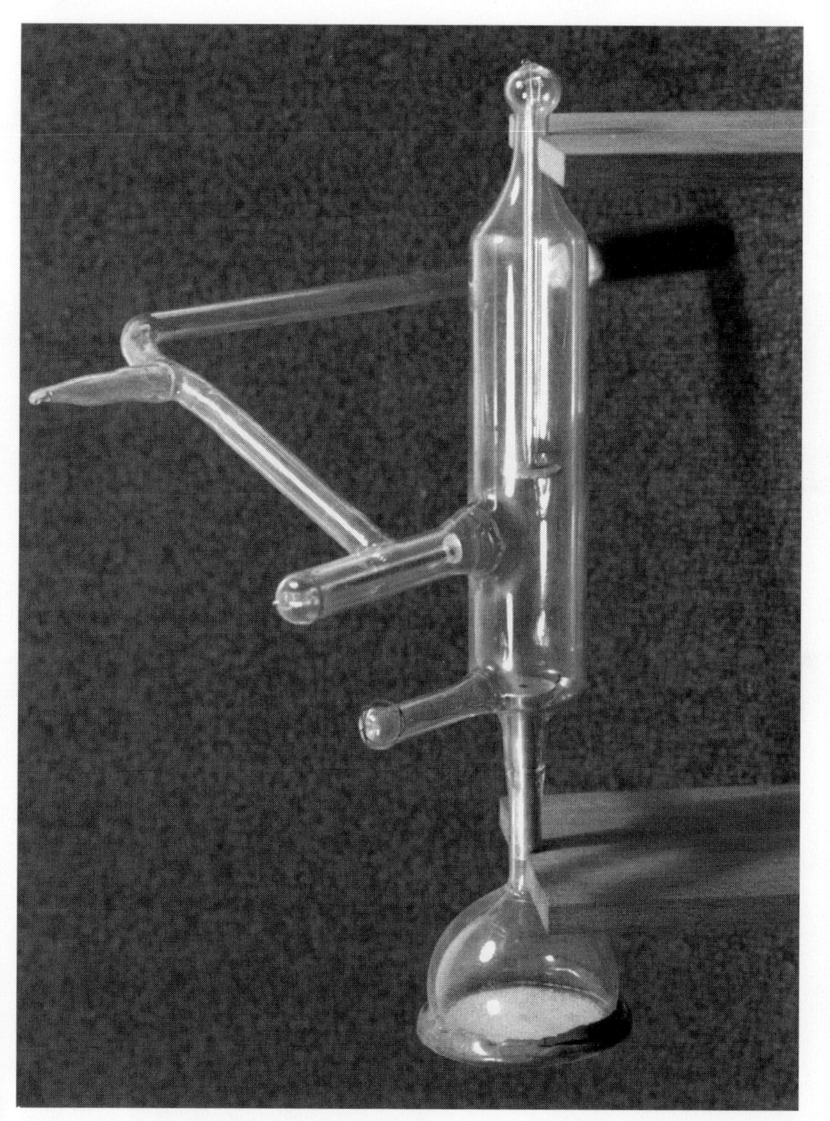

An early positive ray tube, *c.*1907. The positive rays stream to the left through a hole in the cathode (in the narrow section of tube towards the left of the picture). They are detected by fluorescence where they hit the screen on the extreme left. A secondary electrode can be seen in the extension on the top of the tube, which extends further to a tube containing charcoal. This tube was dipped into a flask of liquid air, thus cooling the charcoal which absorbed many of the residual gases in the tube and improved the vacuum. This tube may still be seen in the Cavendish Laboratory Museum

A photographic positive ray tube set up in the laboratory. It has a large discharge bulb, liquid air cooled charcoal evacuation and a magnetic field supplied by a very large electromagnet. The photographic plate is contained in the tall vertical tube on the left

emission is much slower if there are lots of corpuscles in the ring, and Thomson's model demanded hundreds, yet the time still comes when the rings have slowed down, become unstable, and the atom flies apart.

Positive Rays

Thomson's atomic model had great theoretical coherence and promise for unification. Yet it was completely undermined in 1905 by three experiments, suggested by Thomson, which showed that the number of corpuscles in an atom was only comparable with the atomic number. Rather than thousands of corpuscles, atoms contained only a handful. This realisation was fatal to Thomson's theory for two reasons: the mass of the atom must be located in the positive part rather than in the corpuscles, and with only a few corpuscles in them the rotating rings radiated energy too fast to be stable at all. After 1906, Thomson did not publish on the structure of the atom until 1913. The nature of positive electricity was now clearly of burning importance if he wanted to progress further in unifying matter and electricity.

Some years earlier, in 1886, Goldstein had discovered positive rays, or 'Kanalstrahlen' as he called them. These were rays in a discharge tube which streamed backwards through a hole cut in the cathode. By 1905, little was known about them except that they were positively charged and had a charge to mass ratio comparable to that of the hydrogen ion. In them Thomson saw a phenomenon that might be the positive equivalent of cathode rays. In almost exact analogy to his successful cathode ray experiments, he began by investigating the charge to mass ratio of the positive rays by deflecting them in electric and magnetic fields, although this time he applied parallel electric and magnetic fields. This should ensure that particles of a constant charge to mass ratio were deflected into a parabola on the screen at the end of the tube, with a separate parabola for each value of e/m present. Figure 6.2 shows the geometry of the postive ray experiments.

Thomson obtained his results by tracing the patches the positive rays made when they hit a fluorescent screen at the end of the tube. He found that the effects varied greatly with pressure but that the maximum value of e/m was always that of the H^+ ion (10^4) regardless of the gas in the tube and despite precautions to ensure the absence of hydrogen. Moreover, he could not detect any greater fluorescence corresponding to the values of e/m for the other gases in the tube. Reasoning as he had in 1897 for cathode rays, Thomson concluded that the H^+ ion was a fundamental constituent of all atoms.

The fluorescent bands demonstrated the presence of ions with all values of e/m up to the maximum of 10^4. Thomson suggested that these varying values represented a varying charge rather than a varying mass: all the particles they characterized were hydrogen ions constantly combining with, and dissociating from, stray corpuscles in the deflection tube. The value of e/m recorded on the screen amounted to the average charge on the ion during its passage through

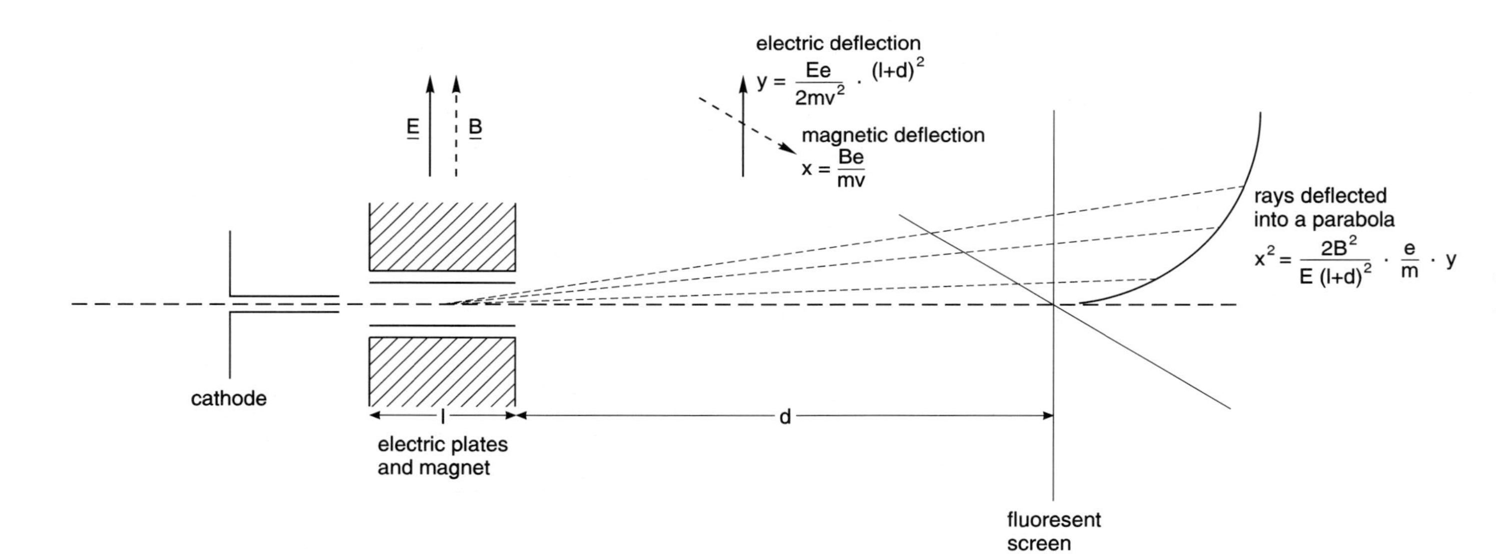

Figure 6.2 The geometry of the positive ray experiments. Positive rays are produced on the left, and stream to the right through the cathode. They are deflected by parallel electric (E) and magnetic (B) fields. Rays of constant e/m are deflected into a parabola on the fluorescent screen on the right: $x^2 = (2B^2/E[l + d]^2)(e/m)y$, where l and d are defined in the diagram, e is the charge on the positive particles, and m is their mass

Thomson measuring positive ray parabolae

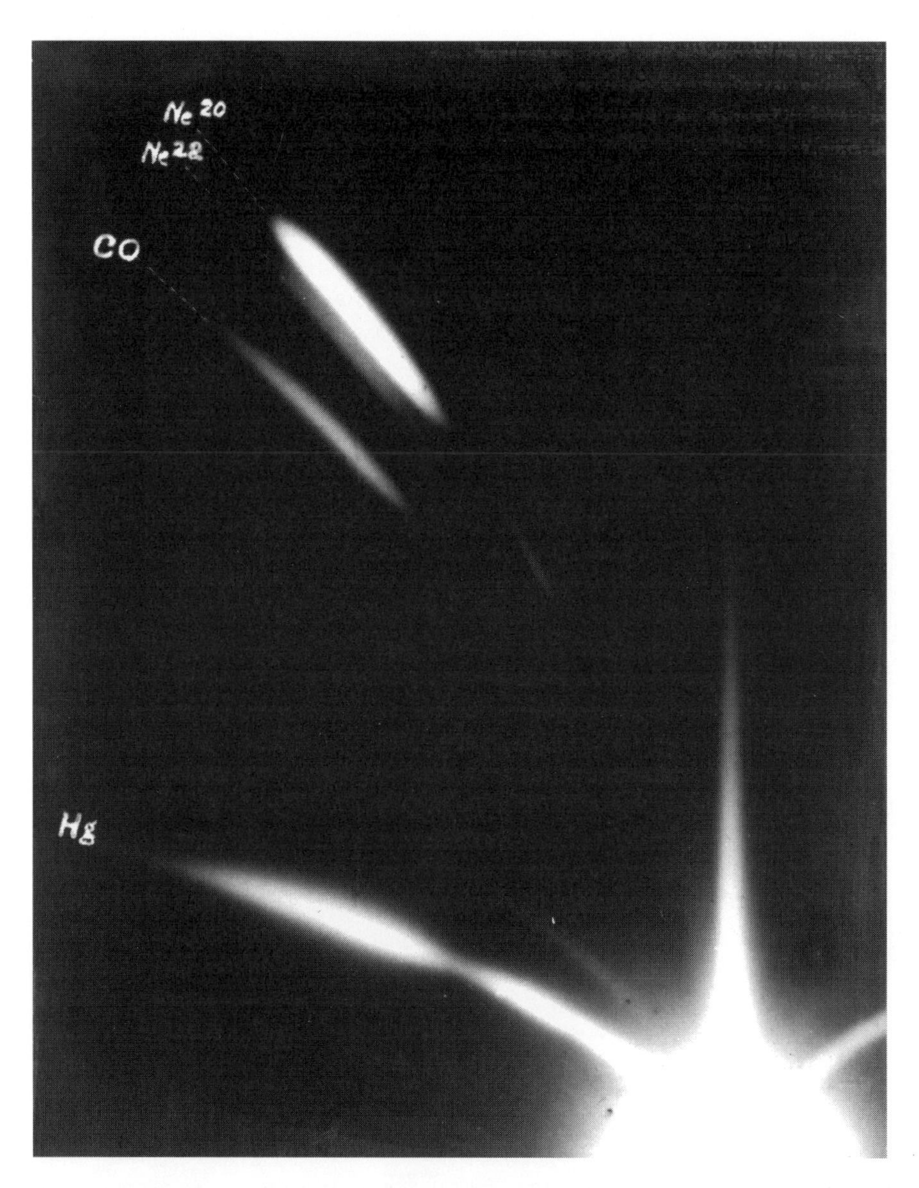

Thomson and Aston's photograph showing parabolae due to the isotopes of neon

the deflection fields. This neutralisation hypothesis followed on from Thomson's theory of discharge by ionisation by collision and subsequent recombination, and recalled his earliest work on discharge where he had postulated the continual pairing and dissociation of vortex rings. It served to maintain his belief that H^+ was a fundamental positive particle.

For the next five years Thomson experimented on the mechanism of neutralisation without any notable success. Then, in 1910, Francis Aston became his research assistant and the experiments took a new turn. Thomson had always realised that he ought to be able to obtain the 'characteristic' parabolae representing all the gases in the discharge tube, but had been unable to do so, and had not pursued the subject. Aston suggested a number of radical changes, use of large-diameter discharge tubes, and isolation of the discharge tube from the deflection tube by use of a very narrow cathode, which enabled the two men to obtain results at very low, previously unobtainable, pressures. They began, for the first time, to see parabolae for all gases in the tube.

A major shock for Thomson came, however, when they started using photographic plates rather than a fluorescent screen, to record their results. He detected many more parabolae than ever before and began to realise that the predominance of H^+ ions in his earlier results was due solely to the insensitivity of his fluorescent screen to heavier ions and not to the universal status of the H^+ ion. Further experiments in 1912, in which he collected and actually counted the ions in each different parabola, showed that the number of H^+ ions was, in fact, extremely small and could easily be due to contamination of the apparatus. He slowly abandoned his idea of the H^+ ion as a fundamental unit of positive charge, instead viewing the entire positive part of the atom as a unit that had not yet been disintegrated artificially, more akin to an atomic nucleus. At the same time he began to allow for the possibility of mechanical as well as electromagnetic mass, writing in some notes for a popular lecture on positive rays: 'May not the positive electricity be like a hermit and of a very stay at home disposition and that in experiments such as those to which I have alluded [the e/m determinations for characteristic rays] we may be weighing the hut plus the hermit, not the hermit himself?' [4]

Although positive rays had proved to tell him little about the nature of positive electricity, Thomson had now found another, though less fundamental use for them as a method of chemical analysis, an aspect that ultimately proved more important.

With their improvements of 1910, Thomson and Aston could observe not only simple ions corresponding to all the gases in the tube, but also multiply charged ions, and molecular ions, some of which represented hitherto unobserved molecular combinations such as the carbon radicals CH, CH_2, CH_3, and the hydroxyl ion OH^+. Thomson's next few papers became largely a list of observed ions.

In this way Thomson and Aston made their most notable discovery with the positive ray tube, that of the isotopes of neon, in 1912. These were the first non-radioactive isotopes to be discovered and, as such, had profound theoretical implications for the development of atomic and nuclear theory. Aston realised their importance and, after the First World War, made it his life's work to separate the two isotopes and put their existence beyond doubt. In doing so he invented the mass spectrometer, a development of the positive ray tube, which is now a standard tool of chemical analysis, and for which he won the Nobel Prize in 1922. Thomson, however, distrusted the isotope interpretation of these neon parabolae which did not fit with his current atomic theory in which isotopes might differ in weight by four units (equivalent to adding an alpha particle to the atomic core) but not less. He was far more excited by another strange isotope they discovered, which he called X_3, and which eventually proved to be an H_3 ion. Thomson's enthusiasm derived from his new theory of chemical constitution, by which the positive core of the atom was built up mainly from tightly bound alpha particles. Except for hydrogen, he observed that the atomic weights of the elements up to 40 fell into two series, one commencing at weight 4 (helium) and increasing in steps of four, and one commencing at 7 (i.e. 3 + 4, lithium) and increasing in steps of four. X_3 might be the first member of this latter series. Thomson's listing of these series in his laboratory notebook is shown in Figure 6.3.

The Structure of Light

In this atomic model of 1913 Thomson attempted, once again, to unify electromagnetism and matter by incorporating his ideas on the structure of light which had been his other dominant concern since 1907. The nature of X-ray ionisation was an outstanding anomaly in his discharge theory. X-rays themselves were still ill understood, though by 1907 the consensus was that they were electromagnetic waves of very short wavelength. But there was a problem with how the rays ionised a gas. They appeared to concentrate their energies on just a few of the gas molecules on which they fell. Why, if the X-rays 'illuminated' the gas evenly, as visible light would, did they ionise only a few of the molecules present?

Thomson's hypothesis, first formed in 1903, that the wave front of X-rays was not in fact uniform, revived his ideas of Faraday tubes. If electromagnetic induction was propagated by vortices in the ether, and if these were finite in number, then a pulse of radiation, X-ray or light, would not be continuously distributed in space. The wave front would be speckled, with bright specks on a dark background, and the energy would be concentrated within the tubes. The limited number of tubes meant that only a limited number of molecules could be ionised.

By 1907, Thomson had recognised a further problem: the energy of the corpuscles emitted on ionisation was too great to have come from the internal energy of the ionised molecule. It must have been absorbed somehow from the incident radiation, and was much higher than one would expect a small atom or molecule to be able to absorb from a uniform, spread out, beam of X-rays.

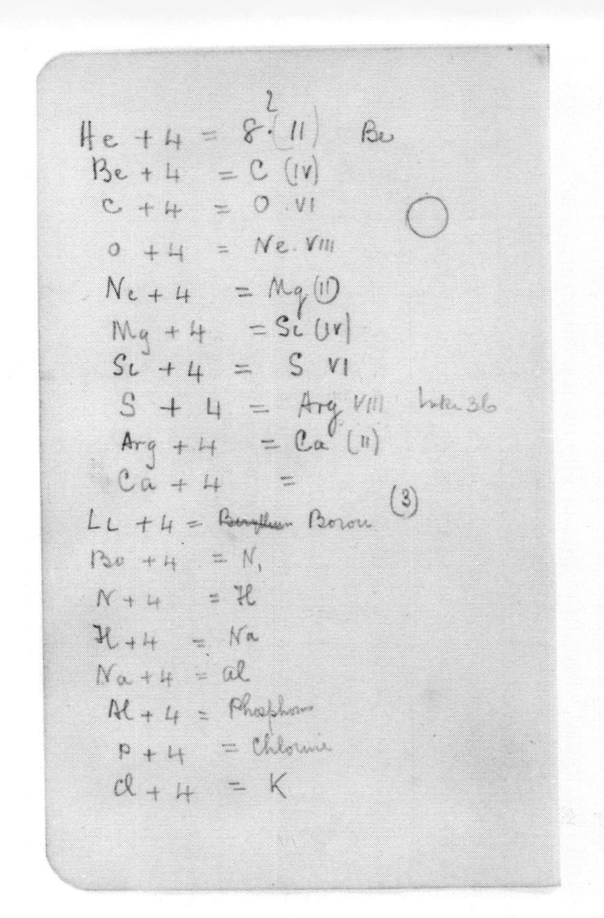

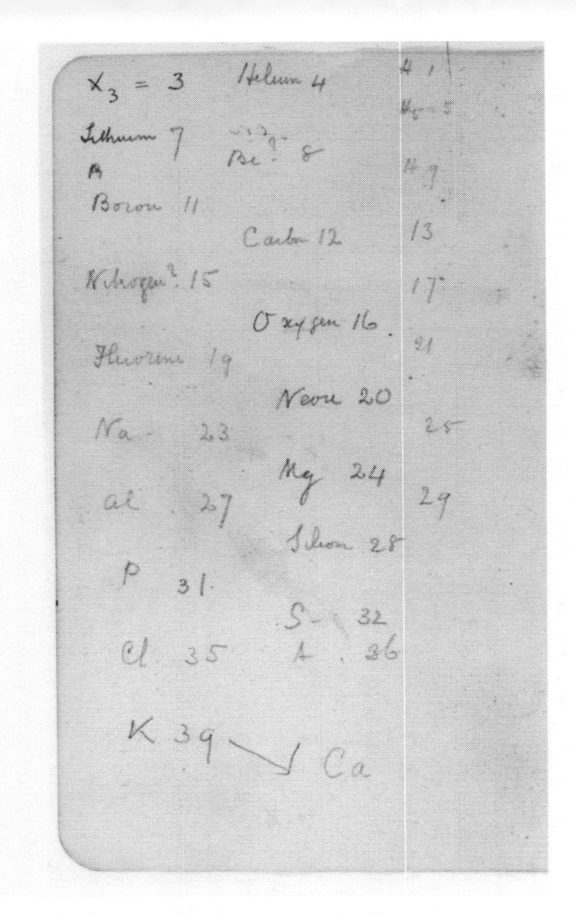

Figure 6.3 Thomson's tables showing the series for the atomic weights of the elements: (left) the $4n$ series above and the $4n + 3$ series below; (right) the $4n + 3$ series on the left, starting with X_3, and the $4n$ series on the right. Courtesy of Cambridge University Library MS, ADD 7654 NB67.

He revived his speckled wave front idea in full force. 'As these units [the Faraday tubes] possess momentum as well as energy they will have all the properties of material particles'. [5] Here he was not suggesting that radiation was particulate, but that it represented an energy structure in a fluid ether. Since he tried to view corpuscles in much the same way, the two, waves and particles, might, he thought, have very similar properties.

Thomson's speculations took place against a background of the long and bitter controversy about the nature of X-rays, wave or particle, which developed between the British William Bragg and Charles Glover Barkla, and also between the Germans Johannes Stark and Arnold Sommerfeld. By 1911 the controversy promoted a crisis in radiation theory which was only finally resolved in the 1920s with the advent of quantum mechanics and wave particle duality.

Meanwhile, Thomson was modifying his Faraday tube idea, looking now at the way X-rays were emitted. In 1910 he suggested that each corpuscle was the origin of a single Faraday tube. When a cathode ray hit the corpuscle, producing X-rays, the corpuscle was displaced suddenly, which resulted in a kink travelling out along the tube (as if a skipping rope is jerked once and the disturbance travels out along it). The kink represented the X-ray, which was thus localised in space.

This idea, however, was difficult to extend to ordinary light because experiments by Thomson's student Geoffrey Ingram Taylor showed that, even at very low intensities, light continued to produce interference effects as if it were an evenly distributed wave. In response, Thomson incorporated into his 1913 atomic model a mechanism for how continuously distributed ultraviolet light might cause the photoelectric effect in which electrons are emitted from metals. The atom, he suggested, consisted of a core, built up in steps of four units of mass, as we have already seen, and surrounded by corpuscles (which he begins to call electrons about now) each one of which was bound in a Faraday tube originating in the core. Incident light caused any electron of appropriate energy to resonate and break out of the tube and then fly away out of the atom because of the repulsion of the remaining electrons. Their energy would be proportional to the frequency of the incident light and Thomson showed that the constant of proportionality was in agreement with Planck's constant. After the First World War he developed this model chemically also, laying the foundations for ideas of covalent bonding.

Views on Physical Theories

By such methods Thomson tried, and continued to try through the 1920s and 1930s, to retain the ether, which most physicists had now abandoned in their adoption of relativity and quantum theory. Thomson considered that these provided merely a mathematical description, devoid of any physical understanding, a subject which he felt strongly about. Of relativity he wrote: 'In

Aston with his mass spectrograph. The similarities to Thomson's positive ray apparatus are clear. The main difference was that electric and magnetic fields were no longer coterminous, and were now perpendicular to each other. This had the effect of focussing ions of differing charge to mass ratio at different points on the photographic plate

J.J. Thomson with his son G.P.Thomson during the First World War

Einstein's theory there is no mention of an ether, but a great deal about space: now space if it is to be of any use in physics must have much the same properties as we ascribe to the ether . . . It would seem that space must possess mass and structure both in time and space: in fact it must possess the qualities postulated for the ether.' [6]

As early as 1893, he had argued for the benefits of 'physical' visualisable theories, such as that of Faraday tubes for 'obtaining rapidly the main features of any problem' before subsequent analysis. [7] Analysis, he said, was 'undoubtedly the greatest thought-saving machine ever invented', adding that 'I confess I do not think it necessary or desirable to use artificial means to prevent students from thinking too much. It frequently happens that more thought is required, and a more vivid idea of the essentials of a problem gained, by a rough solution by a general method, than by a complete solution arrived at by the most recent improvements in the higher analysis.' [8]

Thomson now pointed out that the results explained by quantum mechanics, 'if they were dependent, as they probably were, on universal constants like the velocity of light, the charge and mass of the electron, and Planck's constant, were bound by the theory of dimensions to come out much the same, and that the numerical agreements, such as they were, did not count for much.' [9] As indicated to Thomson years before by analytical dynamics, a large number of theoretical models might produce the same experimental correlations. It behoved the physicist then, to choose the model he found the most productive. In Thomson's case this model was of ethereal vortices.

Last 20 Years

At the outbreak of the First World War, Cambridge was deserted by most undergraduates, and many of the research students joined up. By 1915 the Laboratory was turned over entirely to war work, soldiers were billeted in it, and the workshops were used for making gauges. In July 1915 the Government set up the Board for Invention and Research in an attempt to rationalise the input of scientists to the war effort, hitherto very haphazard. Thomson was one of the four Board members, where a large part of his job seems to have been to smooth the way between the inventors and producers of new equipment, and the eventual users, particularly the Admiralty. Their most successful work was in the development of anti-submarine listening devices. The Board also had to vet ideas put forward by the public, and this provided Thomson with a stock of after-dinner stories for years to come.

> I have just written one letter to a man with a perpetual motion machine, and the other to a charwoman who was much upset by a bad smell and thinks it might be bottled up and used against the Germans. [10]

> Some of the schemes are wild enough, one was to train large numbers of cormorants to peck the mortar from between bricks and then let them loose near Essen so that they might peck the mortar from Krupp's chimneys and so bring them down. [11]

The chairman of the Board was Admiral Lord Fisher, a man notoriously difficult to get on with. Thomson, however, managed extremely well and conceived a great admiration for Fisher. Relations between them were evidently very cordial.

Thomson's tact and ability to get on with people were shown also in his leadership of the Royal Society during these years. He was elected President in 1915 for the usual five-year term. He had previously refused the Presidency because it would take him away from scientific work, but now he was unable to do much experimental science anyway. As President, Thomson acted as the figurehead of science throughout Britain, but he also had to deal with the usual stream of petty squabbles and misunderstandings common to any large organisation. The most serious of these was an attempt to prevent the re-election of Thomson's old friend Arthur Schuster as Secretary because he had been born in Germany and spent his childhood there. Thomson put in a lot of quiet work behind the scenes to ensure the failure of this attempt.

These two offices, member of the Board for Invention and Research and President of the Royal Society, brought Thomson strongly to the attention of the Government as Britain's leading scientist and also a man with a great ability to get on with people. When Dr H.M. Butler, the Master of Trinity College, Cambridge, died in 1917, Thomson was appointed to succeed him. The Mastership is a Crown appointment and his election, the first scientist to hold this premier college headship, not only indicated Thomson's personal prestige, but also established the position of scientists at Cambridge.

Thomson resigned the Cavendish Professorship in 1919 and was succeeded by Rutherford. Thomson and his family moved into Trinity Master's Lodge, where official entertaining became a large part of his role, and he devoted himself to the welfare of the College. He promoted research strongly as of economic benefit to the University and colleges. He was outstanding for his accessibility, welcoming informal contact and discussion with all members of the College, from senior officers to undergraduates. He took a keen interest in student sporting events and enjoyed watching football, cricket and rowing.

He continued as a spokesman for science, being a member of the University Grants Commission from 1919 to 1923, and was instrumental in setting up the Department of Scientific and Industrial Research in 1919. He remained on its advisory council until 1927. His views on science education were much sought, where he argued strongly against too much formalism, and for the positive benefits of research experience which, he thought, encouraged self-reliance, independence of thought and critical awareness, even in those who were not going to make any significant contributions to knowledge.

Thomson pursued experimental work at the Cavendish, as an honorary professor, until a few years before his death. He developed his positive ray experiments further and resumed work on electrodeless discharge, begun in 1890, which contributed to early plasma physics. He published over 50 papers after 1918 and, with his son George, undertook a major revision of his book *Conduction of Electricity Through Gases* in 1933.

In his own work Thomson exemplified those characteristics he admired most in others: enthusiasm and originality. He prized most highly those who were first in any line of work, opening it up to experiment and conjecture, rather than those who claimed to have said the last word. By the end of his life he had come to regard even the search for unification and an ultimate theory as just one more step in a physics which he likened to a divergent series, 'where the terms which are added one after another do not get smaller and smaller, and where the conclusions we draw from the few terms we know cannot be trusted to be those we should draw if further knowledge were at our disposal.' [12]

Thomson died on 30 August 1940, aged 83, having been suffering from ongoing senile decay for the previous two years or so. After cremation, his ashes were buried in the nave of Westminster Abbey, near the graves of Newton, Darwin, Herschel, Kelvin and Rutherford.

Thomson's greatest achievement is generally held to be his identification of cathode rays as electrons, which opened up the whole field of subatomic physics to experimental investigation. But his influence was far wider. His own work marked the transition from nineteenth- to twentieth-century physics. He established one of the world's great research schools. Through his students he set the stage for British physics in the twentieth century, outlining both its interests and methods, in a 'broad brush' approach and a concern for microphysics.

References

Cambridge University Library holds an important collection of Thomson manuscripts, classmark ADD 7654, referred to here as CUL ADD 7654, followed by the particular manuscript number. Other archives, and a more complete bibliography, may be found in: Falconer, I., Theory and Experiment in J.J. Thomson's work on Gaseous Discharge, PhD Thesis, 1985, University of Bath.

[1] THOMSON, J. J., *Recollections and Reflections*, 1936, London: Bell. Reprinted 1975, New York: Arno, pp. 93–94.
[2] Rutherford papers, CUL ADD 7653 T23.
[3] Birmingham University Library, Lodge papers 1/404/18.
[4] CUL ADD 7654 NB65.
[5] THOMSON, J. J., On the ionization of gases by ultra-violet light and on the evidence as to the structure of light afforded by its electrical effects. *Proceedings of the Cambridge Philosophical Society*, **14** (1907) 423.

[6] See reference [1], p. 432.
[7] THOMSON, J. J., *Notes on Recent Researches in Electricity and Magnetism*, 1893, Oxford: Clarendon Press, p. vi.
[8] See reference [7], p. vii.
[9] THOMSON, G. P., *J. J. Thomson and the Cavendish Laboratory in his Day*, 1964, London: Nelson, p. 156.
[10] Letter to G.P. Thomson, 21 July 1915; quoted in Rayleigh, fourth Lord, *The Life of Sir J.J. Thomson*, 1969, London, Dawsons, p. 180.
[11] Letter to H.F. Reid, 5 September 1915; quoted in Rayleigh, fourth Lord, *The Life of Sir J.J. Thomson*, 1969, London, Dawsons, p. 181.
[12] THOMSON, J. J., *The Listener*, 29 January 1930.

Thomson with Rutherford, who succeeded him as Cavendish Professor, taken in the 1930s

Thomson in the 1930s

J.J. Thomson and Rose Thomson in the 1930s

Royal Institution of Great Britain.

WEEKLY EVENING MEETING,

Friday, March 10, 1905.

HIS GRACE THE DUKE OF NORTHUMBERLAND, K.G. D.C.L. F.R.S., President, in the Chair.

PROFESSOR J. J. THOMSON, LL.D. D.Sc. F.R.S., Cavendish Professor of Experimental Physics, University of Cambridge.

The Structure of the Atom.

IN 1897 I had the pleasure of bringing before the Royal Institution experiments showing the existence of *corpuscles*, i. e. negatively electrified bodies having a mass exceedingly small compared with that of an atom of hydrogen, until then the smallest mass recognised in physics. A suggestive and striking property of these corpuscles is that they are always the same from whatever source they may be derived. The corpuscles were first detected in the rays which are projected from the cathode when an electric discharge passes through a vacuum tube, and it was found that whatever the nature of the residual gas in the tube, or whatever the metal used for the electrodes, the corpuscles were always the same. Other sources of corpuscles soon came to light; they were found to be projected from incandescent metals, from metals illuminated by ultra-violet light, and from radio-active substances; but whatever their source the corpuscles were always the same. This fact, in conjunction with their small mass, suggests that these corpuscles form a part of the atom, and my object this evening is to discuss the properties of an atom built up of corpuscles. As these corpuscles are all negatively electrified, they will repel each other, and so if an atom is a collection of corpuscles, there must in addition to the corpuscles be something to hold them together; if the corpuscles form the bricks of the structure, we require mortar to keep them together. I shall suppose that positive electricity acts as the mortar, and that the corpuscles are kept together by the attraction of the positive electricity. We do not know nearly so much about positive as we do about negative electricity; we have never obtained positive electricity associated with masses less than the mass of an atom; in fact, appearances all point to the conclusion that positive electrification is produced by the withdrawal of corpuscles from a previously neutral body. These conditions are satisfied, if we suppose with Lord Kelvin that in the atom we have a sphere uniformly filled with positive electricity, and that the corpuscles are immersed in this sphere. The attraction of the

A

positive electricity will tend to draw the corpuscles to the centre ; the mutual repulsion between the corpuscles will tend to drive them away, and they will arrange themselves so that these tendencies neutralise each other.

Let us now consider the kind of atom we could build up out of corpuscles and positive electricity. The mathematical investigation of this problem leads to the following results. The simplest atom containing 1 corpuscle would have 1 corpuscle at the centre of the sphere of positive electrification ; the 2 corpuscle atom would have the 2 corpuscles separated by a distance equal to the radius of this sphere ; the 3 corpuscle atom would have the 3 corpuscles at the points of an equilateral triangle, whose side is equal to the radius of the sphere ; 4 corpuscles would be at the corners of a regular tetra-

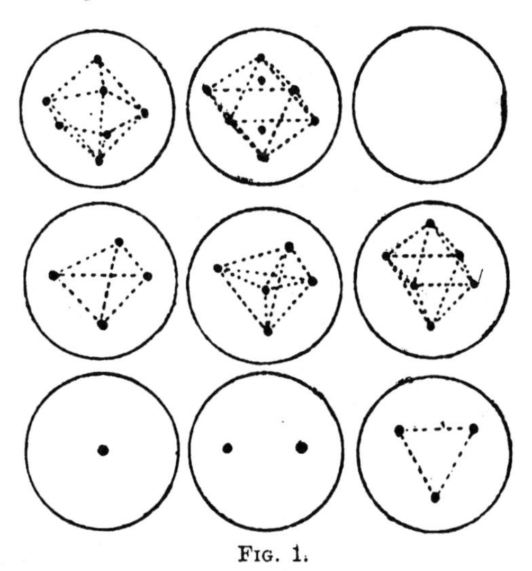

FIG. 1.

hedron, whose side is equal to the radius of the sphere ; 5 corpuscles are situated, 4 at the corners and 1 at the centre of a tetrahedron ; 6 at the corners of an octahedron ; 7 and 8 are more complicated, as the simplest arrangements for 7 and 8, an octahedron with 1 at the centre and a cube, are both unstable ; and for 7 we have a ring of 5 in one plane with 2 on a line through the centre at right angles to the plane ; and 8 we have the octahedron with 2 inside. These arrangements are shown in Fig. 1.

When the number of corpuscles is large, the calculation of the positions of equilibrium becomes very laborious, especially the determination of the stability of the various arrangements. I will therefore treat the subject from an experimental point of view, and apply to this purpose some experiments made with a different object many years ago by an American physicist, Professor Mayer. The problem of the structure of the atom is to find how a number of bodies, which repel each other with forces inversely proportional to the square of the distance between them, will arrange themselves when under the attraction of a force which tends to drag them to a fixed point. In these experiments the corpuscles are replaced by magnetized needles pushed through cork discs and floating on water. These needles having their poles all pointing in the same way repel each other like the corpuscles; the attractive force is due to another magnet placed above the surface of the water, the lower pole of this magnet being of the opposite sign to the upper pole of the floating magnets. This magnet attracts the needles with a force directed to the point on the water surface vertically below the pole of the magnet. The forces acting on the needles are thus analogous to those acting on the corpuscles in our model atom, with the limitation that the needles are constrained to move in one plane.

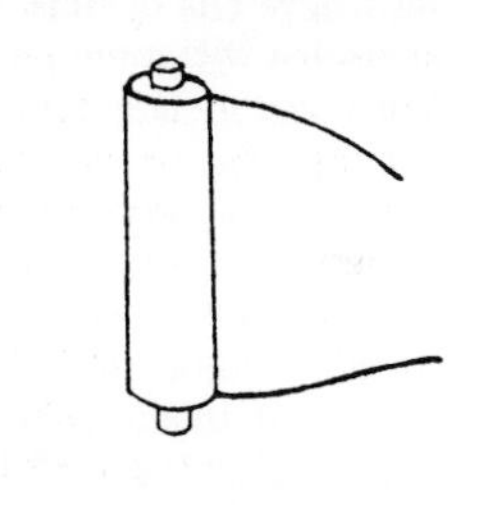

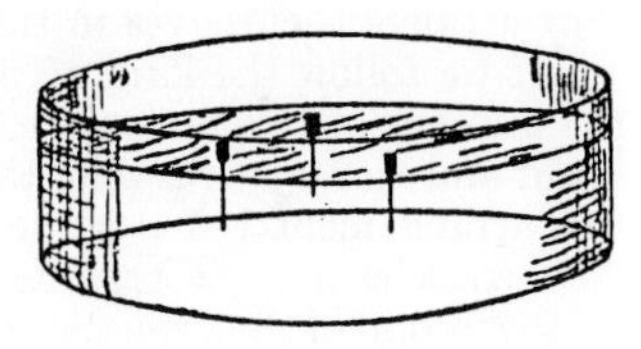

FIG. 2.

As I throw needle after needle into the water you see that they arrange themselves in definite patterns, 3 magnets at the corners of a triangle, 4 at the corners of a square, 5 at the corners of a pentagon; when, however, I throw in the sixth needle this sequence is broken. The 6 needles do not arrange themselves at the corners of a hexagon, but 5 go to the corners of a pentagon, and 1 goes to the middle; a ring of six with none in the inside is unstable. When, however, I throw in a seventh, you see I get the ring of 6 with 1 in the middle; thus a ring of 6, though unstable when hollow, becomes stable as soon as 1 is put in the inside. This is an illustration of the fundamental principle in the architecture of the atom: the structure must be substantial. If you have a certain display of corpuscles on the outside, you must have a corresponding supply in the interior; these atoms cannot have more than a certain proportion of their wares in their windows. If you have a good foundation, however, you can get a large number on the outside. Thus we saw that when the ring was hollow, 5 was the largest number of needles that could be stable. I place in the centre a large bunch of needles and you see that we get an outer ring containing 22 needles in stable equilibrium.

The proportion between the number which is in the outer ring and the number inside required to make the equilibrium stable is shown in the following table:

A 2

Number in outer ring	5	6	7	8	9	10	12	13	15	20	30	40
Number inside. . .	0	1	1	1	2	3	8	10	15	39	101	232

We see from these illustrations how the corpuscles would arrange themselves in the atom, confining ourselves for the present to the case where the corpuscles are constrained to move in one plane. The corpuscles will arrange themselves in a series of rings, the number of corpuscles in the rings getting greater and greater as the radius of the ring gets greater. By the aid of the above table we can readily calculate the way any number of corpuscles will arrange themselves. Let us suppose for example we have 20 corpuscles and try to arrange them so as to have as few rings as possible ; we see from the table that we cannot have more than 12 in the outside ring, for 13 would require 10 inside, and would be impossible with less than 23 corpuscles : thus 12 will be the number in the outside ring and there are eight left to dispose of ; these cannot form a single ring with no corpuscles inside, as 5 is the greatest number that can do this ; the 8 corpuscles will therefore break up into two systems, a ring of 7 with 1 inside. You see that when I try the experiment with 20 magnets they arrange themselves in this way.

If we follow the kind of atoms produced as we gradually increase the number of corpuscles, we find that certain arrangements will recur again and again ; thus take the case of 20 corpuscles ; this consists of the arrangement 1–7–12, the arrangement for 8 is 1–7 ; the atom of 20 corpuscles may be regarded as formed by putting another storey to the atom of 8 corpuscles ; if we go to 37 corpuscles, we find the arrangement is 1–7–12–17, i.e. another storey added to the atom of 20, while for 56 we have 1–7–12–17–19, the atom of 37 with another storey added. Thus the possible atoms formed by numbers of corpuscles from one to infinity could be arranged in classes, in which each member of the class is formed by adding another storey to the preceding member ; the structures of all the atoms in this class have much in common, and we might therefore expect the physical as well as the chemical properties of the atoms to have a general resemblance to each other. This property is, I think, analogous to that indicated by the periodic law in chemistry. We know that if we arrange the elements in the order of their atomic weights, then, as we proceed in the direction of increasing atomic weight, we come across an element, say lithium, with a certain property ; we go on and after passing many elements which do not resemble lithium, we come across another, sodium, having many qualities in common with lithium. Then as we go on, we lose these properties and come across them again when we arrive at potassium ; exactly the kind of recurrence we should get with our model atoms, if we suppose the number of corpuscles in the atom to be determined by its atomic weight.

Let me give another instance of the way the properties of these atoms resemble the properties of the chemical atom. I will take the electro-chemical property of the atom. Some atoms, such as those of lithium, sodium, potassium, have a strong tendency to be positively electrified, while others like chlorine, bromine, iodine, tend to be negatively electrified. Now the way our model atom gets positively electrified, is by losing a negatively electrified corpuscle ; thus, those atoms in which the corpuscles are loosely held would tend to get positively electrified, while those whose corpuscles are very firmly held would not get positively electrified, and might be able to bear the disturbance due to another corpuscle placed outside without disintegration, and with this additional corpuscle they would be negatively charged. Now let us see how this property would vary from atom to atom. I will take a numerical case. Suppose we begin with 59 corpuscles ; we should have by the table 20 on the outside, and 39 in the inside ; but as 39 is the least number of corpuscles that can hold a ring of 20 in stable equilibrium, the equilibrium of this atom would have nothing to spare ; it would be in rather a tottery condition, and a corpuscle would be easily detached, leaving the atom positively charged. Let us now go to the atom with 60 corpuscles ; it would still have 20 on the outside, but it would have 40 on the inside, and be more stable than 59 ; it would not so easily lose a corpuscle ; and would not thus be so electro-positive as 59 ; as we go on up to 67 we have still 20 on the outside but get more and more in the inside, the difficulty of getting a corpuscle out therefore increasing, and the atom getting more and more electro-negative. Let us see what happens when we get to 68 ; here we have 21 on the outside and 47 inside, but as 47 is the smallest number which can keep 21 in equilibrium, this equilibrium is shaky, and as in the case of 59 corpuscles the atom would be very electro-positive. Thus, as we increase the atomic weight, we get for a certain range, a continual diminution in the electro-positive character ; this goes on until we get to 67, then there is a sudden jump from the electro-negative 67 to the electro-positive 68, followed again for a time by a continual decrease in electro-positive characteristics with increasing atomic weight. Compare this with the behaviour of the atoms of the chemical elements

Li	Bi	Bo	C	N	O	Fl
Na	Mg	Al	Si	P	S	Cl
K	—	—	—	—	—	—

The electro-positive character diminishes as we proceed from Li to Fl, then there is a sudden change from the electro-negative Fl to the electro-positive Na, then another diminution in the electro-positive character to Cl, and then another sudden change from Cl to K.

The model atoms we are considering are all built up of the same materials—positive electricity and corpuscles—hence the atoms of any one element would furnish the raw materials for the atoms of any other element, and a rearrangement of the positive electricity and corpuscles would produce transmutation of the elements. Whether the atoms of our elements will tend to break up into the atoms of other elements will depend upon the relative stability of the atoms, and the stability of an atom will depend mainly upon its potential energy; if this is large, the atom will be liable to break up or change. I have calculated for atoms containing from 1 to 8 corpuscles the potential energy of the atom per corpuscle: i.e. the potential energy of the atom divided by the number of corpuscles in the atom, making the assumption that positive electricity behaves like an incompressible fluid, i.e. that its density is invariable. The result is represented graphically in Fig. 3; the vertical ordinates represent the potential energy per corpuscle, the horizontal abscissæ the number of corpuscles in the atom. You will notice that the curve is a wavy line with peaks and valleys; the atoms corresponding to the peaks would have greater potential energy than their neighbours, and would therefore tend to be unstable, while those in the valleys, having relatively little potential energy, would be stable.

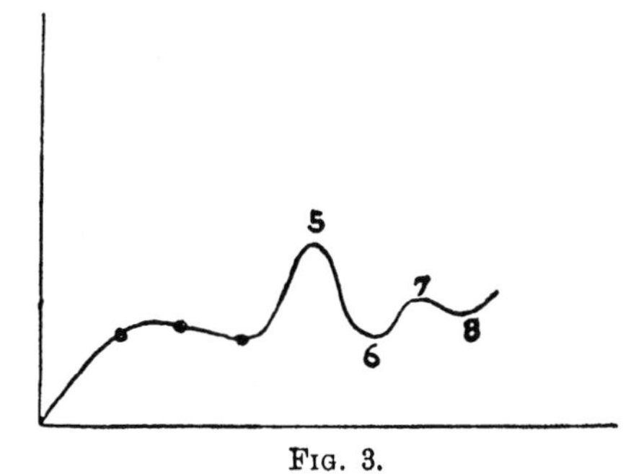

Fig. 3.

The case is in many respects very analogous to the case of a number of stones scattered over a hilly country whose section is represented by Fig. 3; the stones, if subject to disturbances, would run from the hills into the valleys, and though the stones might be uniformly distributed to begin with, yet in course of time they would accumulate in the valleys. So also in the chemical problem, though the number of atoms of the different elements might initially not be very unequal, yet, in course of time, those in the valleys would increase, and those on the peaks diminish, so that some elements would increase, while others would tend to become extinct. The smallest potential energy is that of an atom consisting of a single corpuscle; this is the goal which all the atoms would ultimately reach, if subject to disturbances sufficiently intense to lift them over the intervening peaks. Thus, on this view, the general trend of the universe would be towards simplification of the atom—though there might be local eddies. The final stage would be that in which all the atoms contained only one corpuscle. This result depends upon the assumption that the positive electricity is incompressible, i.e. that its density is constant; if we had assumed that the volume of the positive electrification is the same whatever may be the quantity of electricity, we should have found that, although there would still have been changes from one element to another, the general trend would have been in the opposite direction, i.e. the simple atoms containing only one corpuscle would gradually condense into more and more complex atoms.

Chemical Combination. Action of the Atoms on each other.—We have hitherto confined our attention to the consideration of the stability of the arrangements of the corpuscles in the atom. We shall now proceed to discuss the question of the action of one atom on another, and the possibility of the existence of stable configurations of several atoms, in fact the problem of chemical combination.

As far as I know, the only cases in which the conditions for equilibrium or stable steady motion of several bodies acting upon each other have been investigated, is that suggested by the solar system; the case in which a number of bodies—suns, planets, satellites—attract each other with forces inversely proportional to the square of the distance between them. The complete solution of this problem, or anything approaching a complete solution, has proved to be beyond the powers of our mathematical analysis; but enough has been done to show that with this law of force, stable arrangements of the mutually attracting bodies only occur under stringent conditions. Thus, to take a very simple case, that of three bodies, it has been shown that, when the bodies are equal, there is no arrangement in which the steady motion is stable; if, however, the masses are very unequal, then it is possible for such an arrangement to exist. Another very interesting case is one investigated by Maxwell in connection with the theory of Saturn's rings. It is that of a large planet surrounded by a ring of satellites, each satellite following its neighbour at equal intervals round one circular orbit. Maxwell showed that this system was only stable under certain conditions, the most important being that the mass of the planet must be much greater than that of the satellite. The proportion between the mass of the smallest planet able to retain the ring in steady motion and the mass of one of the satellites increases very rapidly as the number of the satellites increases: if P is the mass of the planet, S that of a satellite, n the number of satellites, Maxwell showed P must be greater than $\cdot 43\, n^3$ S. The consequences of this are interesting from the analogy shown in the case of chemical combination. Thus, suppose the mass of a satellite were $\frac{1}{100}$ part of that of the planet, then the result shows that the planet could retain 1, 2, 3, 4, 5, 6 satellites, but not more than 6. With 6 satellites the planet is, to use a chemical term, saturated with satellites, and the behaviour of the system is equivalent to that of the atom of a sexavalent element, which can unite with 6 but with not more than 6 atoms of hydrogen.

The existence of a limit to the number of systems in a ring, which a central system can hold in stable equilibrium, is not peculiar to any special law of force. We have already seen examples of it inside the atom, where the central force on the satellites is supposed to be proportional to the distance. We have just seen that it holds in the planetary system, where the central force varies inversely as the square of the distance. I have found that this limit exists for all the laws of force I have tried, although of course the number of satellites which can be held in equilibrium depends on, among other things, the law of force.

The law of the inverse square is not favourable to the formation of stable systems, even when, as in the astronomical problem, the forces between the various bodies are all attractive; it is quite inconsistent with stability when, as in the case of the chemical atoms, some of these bodies carry charges of the same sign, and so repel each other. Thus, suppose we have the central body charged with positive electricity, while the satellites are all negatively electrified, so that the central body attracts the satellites, while the satellites repel each other. With forces varying inversely as the square of the distances between them, it is easy to show that with more than one satellite stability is impossible.

The mathematical investigation of the case where the satellites repel each other shows that, in order to ensure stability, the central attraction must, in the neighbourhood of the satellite, increase when the distance of the satellite from the planet increases. Inside the atom we have supposed that the central attraction was proportional to the distance from the centre, so that in this region the central force increases rapidly with the distance at all points. It is not necessary for equilibrium that the increase should be as rapid as this, nor indeed that the force should everywhere increase with the distance; all that is necessary is that in the neighbourhood of the satellite the force should increase and not decrease as the distance increases.

It might appear at the outset as though atoms of the kind we have been considering, made up of positive electricity and corpuscles, could never form stable arrangements, for there is a theorem known as Earnshaw's theorem, to the effect that a system of bodies attracting or repelling each other with forces varying inversely as the square of the distance between them, cannot be in stable equilibrium. This result does not prevent the existence of stable arrangement of atoms in the molecule, for Earnshaw's theorem only applies to the case when the bodies are at rest; it does not preclude the existence of a state of steady motion, in which there is no relative motion of the atoms. Again, in the case of our atoms there are other forces besides the electrostatic attractions and repulsions, for if the corpuscles are in rotation inside the atom, they will produce magnetic forces, so that outside the atom there will be a magnetic, as well as an electric field. The magnetic field will greatly promote the stability of the atoms if these are charged, for it will, if strong, practically prevent motion at right angles to the direction of the magnetic force, so that the arrangement of atoms will be stable provided the electrostatic forces give stability for displacements *along* the lines of magnetic force. For example, if at any point near an atom the magnetic force were radial, then a second charged atom at this point would be in stable equilibrium, provided the radial attraction between the atoms at that point increased as the distance between the atoms increased.

Let us now consider the forces produced by an atom of the kind we have described. Take the case of an uncharged atom, i.e. one where the sum of the charges on the negatively electrified corpuscles is just equal to the positive charge in the sphere in which the corpuscles are supposed to be placed. Let us consider the radial force to the centre due to such an atom. Since there is as much positive as negative electricity in the atom, the average radial force taken over the surface of a sphere with its centre at the atom is zero ; this does not mean that the radial force is everywhere zero, but that at some places it is directed towards the centre, and at others away from it. There

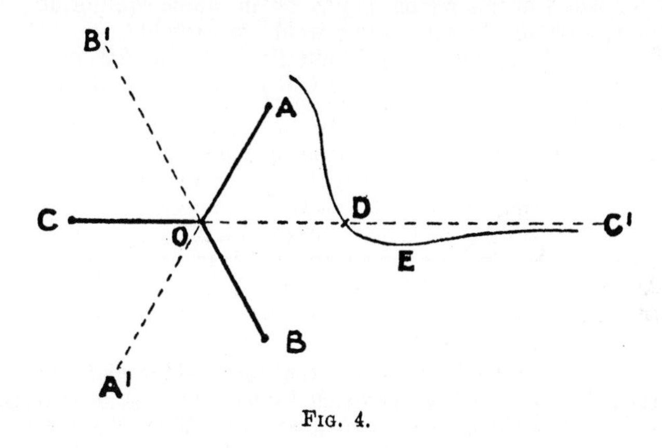

Fig. 4.

may be, as we shall see, certain directions in which the force changes from attraction to repulsion, or *vice-versâ*, as we travel outwards from the sphere.

Thus take the case of three corpuscles placed in a sphere. The corpuscles, when in equilibrium, are at the corners of an equilateral triangle A B C ; let O be the centre of the atoms of which these corpuscles form a part. Consider the force on a positively charged particle. As we travel from A radially outwards, we find that the force is always towards O, and gets smaller and smaller as we get further and further away. As the attraction diminishes as the distance increases, there is no place at which the particle would be in equilibrium, stable or unstable. Suppose, however, we travel outwards along O C^1, the prolongation of C O, then when the particle is just outside A B, the force on the particle is repulsive. This repulsive force diminishes as we recede from the atom and vanishes at a certain distance D ; at greater distances from the atom than D, the force is attractive and remains attractive at all greater distances; thus a positively charged particle would be in equilibrium at D, and it is easy to see that the equilibrium would be stable, for if the particle were made to approach O, the repulsive force would drive it back to D, while, if the particle were to recede from D, the attractive force would drag it back. If we represent the relation between the radial force and the distance by a graph, a point above the horizontal axis corresponding to repulsion, and one below it to attraction, we obtain a curve of the following character. The curve crosses the axis at the point D, the place where the force vanishes; after passing D, the force which is now attractive increases as the distance from the atom increases, until a point E is reached when the force is a maximum; beyond E the attraction diminishes as the distance increases. Thus, since in the region D E, the force is attractive and increases as the distance increases, a positive particle, placed in this region, might be in stable equilibrium, while outside this region the equilibrium would be unstable.

There would, of course, by symmetry be similar regions on O A¹, O B¹, the prolongations of O A and O B respectively. It will be seen

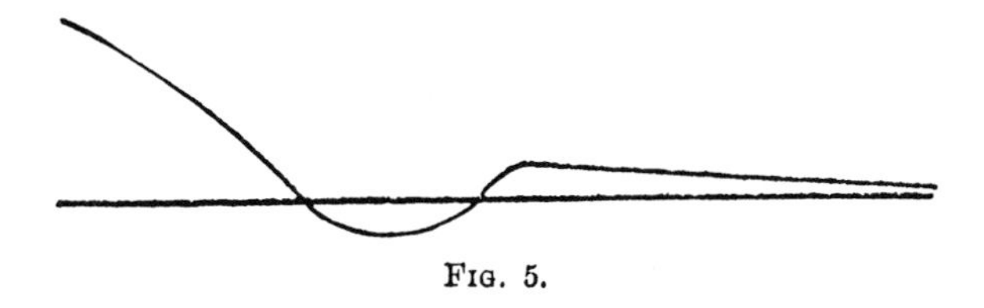

Fig. 5.

that the nature of the force betwen the atom and the charged particle, is of the type postulated by Boscovich, i.e. a repulsion at short distances succeeded by an attraction at greater ones. With the very simple type of atom we have been discussing, there is only one change from repulsion to attraction; with atoms containing more corpuscles, the graph representing the relation between force and distance becomes more complicated, and we may have several alternations between repulsion and attraction instead of only one as in Fig. 5.

However complicated the atom, a distribution of forces of this kind will only occur in a limited number of directions, or rather only along directions making small angles with a limited number of axes drawn in definite directions.

I have here an arrangement to show the change in direction of the force due to an atom. The atom is supposed to be one with three corpuscles; these are represented by the negative ends of three electromagnets arranged radially on a board, the positive ends of the magnets which represent the positive electrification in the sphere being at the centre. We see that along the lines O A¹, O B¹, O C¹, the magnetic force on a positive pole changes from repulsion to attraction at a certain distance, and that the system can hold three floating magnets in stable equilibrium at a finite distance from its centre.

An atom analogous to the one we have just been considering would have the power of keeping three positively electrified particles

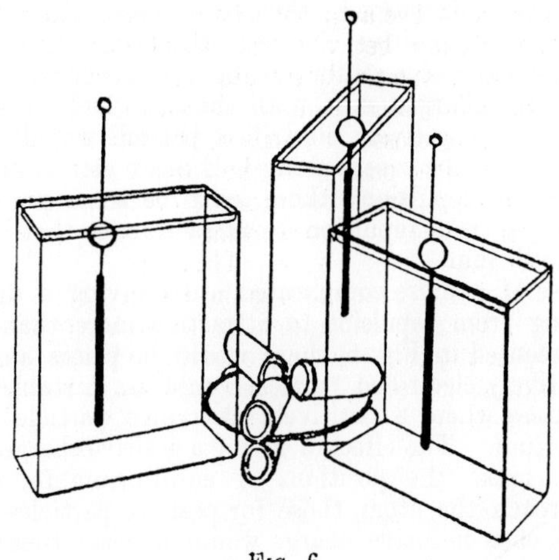

Fig. 6.

in stable equilibrium, provided these are placed at suitable distances along the lines O A¹, O B¹, O C¹. With other arrangements of corpuscles, we should get atoms able to keep negatively electrified particles in equilibrium. Thus, for example, if we have 5 corpuscles placed at the corners of a double pyramid as in Fig. 7, then along the lines OA, O B, O C, at suitable distances from O negatively electrified particles could be in equilibrium, even if the atom were uncharged. If, however, the central atom were uncharged while the satellites were charged, the molecule, as a whole, would be charged, whereas we know the molecule is electrically neutral ; we must consider, therefore, what would be the effect of giving a charge of electricity to the central atom.

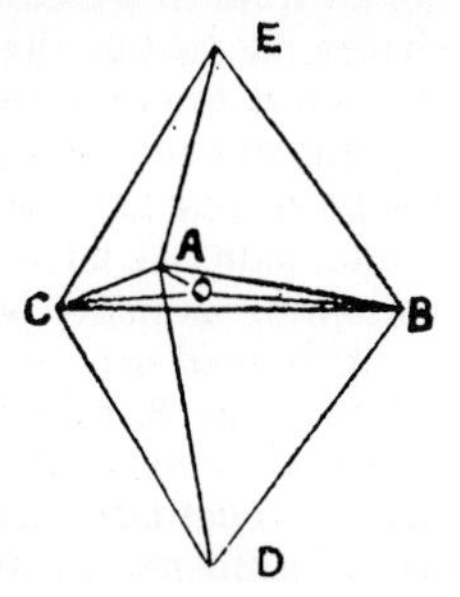

Fig. 7.

In the case of the three corpuscles, if we gave a negative charge to the central atom, the axes O A¹, O B¹, O C¹, might or might not cease to be axes of stable equilibrium for positively electrified particles. The effect of the charge would be to bring the point D of equilibrium closer to the atom—how much closer would depend upon the charge given to the atom ; but as long as D kept outside the atom, stable equilibrium for positively electrified particles would be possible ; if, however, D came inside the atom, the axes O A[1], O B[1], O C[1], would cease to be axes of possible equilibrium.

In some cases, the communication of a charge to the atom might, in addition to affecting the position of equilibrium along the axes for the uncharged atom, introduce axes of stability which did not exist when the atom was uncharged ; thus, in the case of a double pyramid Fig. 7, if we gave a positive charge to the atom, the axes O E, O D, which were not axes of equilibrium for the uncharged atom, would become so for the charged one ; for if the atom had a positive charge, the force on the negatively electrified particle would at a point a great distance from the centre along O E be an attraction, while close to E it would be a repulsion ; there must be some point then when the force changes from repulsion to attraction, so that this axis will be one of equilibrium.

In the case of a more complicated atom giving a distribution of force changing from repulsion to attraction more than once, as in the case represented in Fig. 5, there would be places along this axis where a negatively electrified particle would be in stable equilibrium and other places where a positively electrified particle would be in stable equilibrium. The effect of giving a positive charge to this alone would be to make the positions of equilibrium for the negative particles approach the atom, those for positive particles recede from it ; the effect of a negative charge would displace those positions in the opposite directions.

The forces we have been considering are those exerted by an atom on a charged particle ; they would be a part (and in many cases, I think, the most important part) of the forces acting on a second atom, if that atom had an excess of one kind of electricity over the other. Remembering, however, that there is an electric field round an atom, even when it is uncharged, and that an uncharged atom is not an atom in which there is no electricity, but one where the negative charge is equal to the positive, we easily see that two uncharged atoms may exert forces on each other ; the calculation of these forces is, on account of the complex nature of the atom, very intricate, and I shall not go into it this evening. I shall treat the subject from the experimental side. I have here two systems, each built up of magnets, each containing as many positive as negative poles, and thus analogous to an uncharged atom ; one of them is suspended from the arm of a balance, Fig. 8. You see that I can place these systems so that they repel each other when close together and attract each other when further apart, so that these atoms would be in stable equilibrium under each other's influence when separated by the distance at which repulsion changes to attraction.

The force which an atom A exerts on another atom B may be conveniently divided into two parts : the first part, which we shall call the force of the E type, depends upon the charge on B ; it is proportional to this charge and independent of the structure of B, and we might, without altering this force, replace B by any atom we pleased, provided it carried the same charge. The other part of the force, which we shall call the M part, is independent of the charge on B, but depends essentially on its structure; this part of the force would be entirely altered, if we replaced B by an atom of a different kind.

The question now arises, What part do these two types of force play in determining the nature of the molecule? Is the stability determined by forces of the E or of the M type?

The E forces depend on the charges carried by the atom, so that in those compounds in which stability is due to the E forces, the

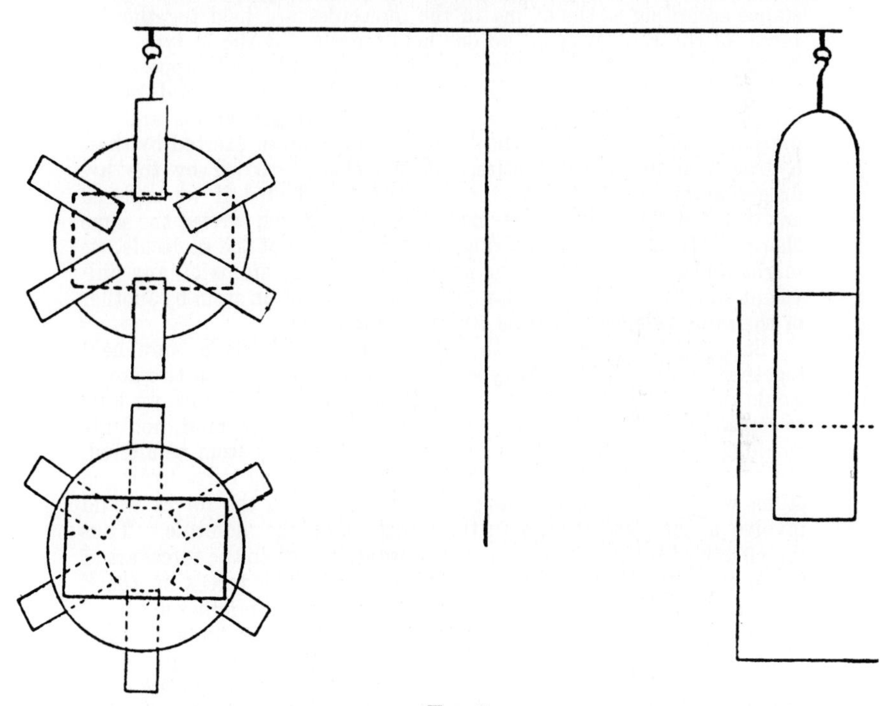

Fig. 8.

atoms must be charged. We are thus confronted with the question, Are the atoms in a molecule charged with electricity, or are they electrically neutral? Thus, to take a definite case, in the molecule of marsh gas, which we picture as a carbon atom at the centre of a tetrahedron with the four hydrogen atoms at the corners, are the hydrogen atoms charged with equal quantities of negative electricity, the carbon atom having a four-fold charge of positive, or are both carbon and hydrogen atoms uncharged? It is difficult to get direct evidence on this point, since the molecule as a whole is neutral on either supposition. There is, however, considerable indirect evidence to support the view that the atoms in many compounds are electrified. I may mention, as examples of such evidence, the power possessed by certain molecules, such as those of sugar, of rotating the plane of polarisation of light passing through them,. This power, which is associated with the presence of the asymmetric carbon atom with four dissimilar atoms attached to it, is readily explained by the electromagnetic theory of light; if the atoms in the molecule are charged, it is difficult to see how uncharged atoms could produce sufficient rotation.

Let us consider the difference in the chemical properties of a substance according as the atoms in the molecules are held together by forces of the E or M type and one held together by the M type. Let us take the molecule of marsh gas as an example, and suppose that the molecule is in equilibrium under the E forces exerted by the carbon atom on the negatively electrified hydrogen atoms and the mutual repulsions between these atoms. The forces exerted by these hydrogen atoms depend entirely on the charge carried by the hydrogen atom; none of these forces would be affected if we replaced any or all of the hydrogen atoms by any atom which carried the same charge. Hence, without altering the architecture of the molecule, we might replace any or all of the hydrogen atoms by atoms of any univalent substance. In this case, the replacement of an atom by another of the same valency would be a very simple thing.

Suppose, however, that the atoms in the molecule were held together by forces of the M type, then the forces between two atoms would depend on the structure of both the atoms. If now we were to replace one of the H atoms by an atom of another kind, not only would the force exerted by the carbon atom on this atom be altered, but the forces exerted by the atoms on the remaining three hydrogen atoms would be radically changed; this change in the forces would involve a complete change in the structure of the molecule. Thus the effects of replacement are much more serious when the forces are of the M type than when they are of the E type. The forces of the E type are, I think, those which are most effective in binding atoms of different kinds together, while the M type of forces finds its chief scope in binding similar atoms together as in the molecule of an element, or as in the connecting the carbon atoms in the carbon compounds.

Let us sum up the results we have arrived at. We have seen that an atom built up of corpuscles in the way we have described possesses, whether charged or uncharged, the following properties. There are certain directions fixed in the atoms, along which or in directions not too remote from which, electrified particles, positively electrified for some kinds of atoms, negatively electrified for others, and either positively or negatively electrified for still other kinds of atoms, will be in stable equilibrium, if placed at suitable distances from the centre of the atom. We may call those directions the valency directions, and the regions within which the equilibrium is stable the valency regions. Those who are familiar with the beautiful theory of Van't Hoff and

Le Bel on the asymmetric carbon atom, which supposes that the attractions exerted by a carbon atom are exerted in certain definite directions, these directions being such that, if the carbon atom is at the centre of a regular tetrahedron, the attractions are along the lines drawn from the centre to the corners, will perceive the resemblance between that theory and the results we have been discussing. There is, however, an important difference between the two, for on our theory the forces exerted by the atom are not confined to any special direction; the atom exerts forces all round. It is only, however, in certain directions that these forces can keep a second atom in stable equilibrium. We picture, then, the atom A as being connected with a limited number of closed regions of finite size, and any body attached to the atom must be situated in one of these regions; when each of these regions is occupied by another atom, the atom A can hold no more bound to it, and is said to be saturated.

I have not time this evening to discuss in any detail further developments of these ideas. I may however, in conclusion, call attention to a point which is illustrated by the behaviour of the carbon compounds. Suppose that C_1 C_2 are two carbon atoms near together. Then when

$$C_1 \quad \alpha \quad \beta \quad C_2$$

both atoms are present, regions α, β near the line joining C_1 C_2, which were valency regions for C_1 and C_2 when these atoms were alone, may cease to be valency regions when both are present. For take the case when the stability is due to the magnetic force produced by the rotation of the corpuscles within the atoms. Along the line C_1 C_2, the magnetic force due to C_1 and C_2 will be in opposite directions, and in the region near the middle of C_1 C_2 the resultant magnetic force would be very small, so that in this the equilibrium of a charged body would be unstable; thus α β would cease to be valency regions. This reasoning would not apply to the valency regions of C_1 on the side opposite to C_2, nor of those of C_2 on the side away from C_1, so that six valency regions would remain. Thus if we consider the tetrahedra formed by the valency regions round our carbon atoms, then if two carbon atoms are placed so that two vertices of these tetrahedra come together, the regions near these vertices will cease to be valency regions, and the compound formed would have to be of the type $\begin{matrix} H \diagdown \\ H - C \\ H \diagup \end{matrix} \; \begin{matrix} \diagup H \\ C - H \\ \diagdown H \end{matrix}$, the two carbon atoms being held together by forces of the M type. If the tetrahedra were placed so that two edges of the tetrahedra came together, we could show similarly that the four valency regions at the ends of the edge would be suppressed and the compound would be of the type, $\begin{matrix} H \\ H \end{matrix}\!>\!C \; C\!<\!\begin{matrix} H \\ H \end{matrix}$, while if two faces of the tetrahedra came together the valency regions in these faces would be suppressed, and the compound would be of the type H—C C—H.

7

Subsequent Developments

Two threads of scientific development – one fundamental, the other technological – can be traced back to the discovery of the electron. The first of these, namely the search for elementary particles (including an understanding of their properties and interactions), grew from modest beginnings in Thomson's evacuated tubes into a major research field requiring large and expensive particle accelerators. The second, the 'electronics revolution', has seen the invention and development of a variety of devices, notably the valve, transistor and integrated circuits, which have transformed both science and society via their applications in instrumentation, communication, television and electronic computers.

Elementary Particle Physics

Speculations concerning the ultimate divisibility of matter date back to the fifth century BC when the Greek philosopher, Democritus, used the term *atom* to denote the smallest component of which substances are composed. The Roman poet, Lucretius eloquently expressed the view that atoms, along with the empty space, are the only eternal and immutable entities of which our physical world is made. Little changed between these speculative musings and the beginning of the nineteenth century, when the English scientist, Dalton laid down the foundations of modern chemistry with his theory of the elements. The view that all material substances were composed of indivisible atoms (with the variety of known matter resulting from the way in which they combined) then became sacrosanct. Almost one hundred years later, Thomson's demonstration that atoms themselves were divisible into smaller particles destroyed the fundamental status of the atom and undermined belief in the indivisibility of any particle. The search began for other particles and for knowledge of their structure.

As we have seen, although Thomson initially regarded his negatively charged corpuscles (electrons) as the sole constituents of atoms, he recognised the need for compensating positive charge to preserve neutrality. His model of the atom in which electrons were immersed in a positively charged paste like 'the plums in a pudding' was repudiated by Rutherford's famous scattering experiments which showed that the positive charge, and most of the mass of the atom, was concentrated in a tiny nucleus at its centre. Rutherford gave the name *proton* to the nucleus of hydrogen, the lightest atom, and in 1914 Niels Bohr proposed a model of this element which consisted of a single electron circling the proton, like a planet around the sun, but with an attractive electrostatic rather than a gravitational force maintaining the orbit.

Extension of this picture to the heavier elements encountered a problem, namely that while it seemed reasonable to suppose the nuclei of these contained two, three or more protons with an equal number of orbiting electrons, the masses of the elements were not as expected according to this model. Helium, with two electrons and two protons, is four times as massive as hydrogen; lithium with three electrons and protons is seven times as massive, and so on. This dilemma was solved by James Chadwick with his discovery in 1932 of a third fundamental particle, the *neutron*, which carries no net charge but which has a mass essentially equal to that of the proton. Neutrons share the nucleus with protons and make up the mass difference; so, for example, the nucleus of a helium atom contains two protons and two neutrons, while the lithium nucleus has three protons and four neutrons. In fact, the nuclei of all elements, apart from hydrogen, contain both protons and neutrons.

With the discovery of the neutron, the question 'What is matter made of?' appeared to have been answered in a very simple and satisfying manner and indeed, for many purposes today three fundamental particles – electrons, protons and neutrons – suffice to explain most of the important properties of materials. However, the simplicity of what is sometimes called 'the classical model' of atoms was shattered when, just over thirty years ago, it was discovered that protons and neutrons were not in fact indivisible entities but were themselves made up of smaller particles which were named *quarks*.[1] These new particles were identified in experiments in which electrons were arranged to collide with protrons or electrons at very high energies in particle accelerators. There are six types or 'flavours' of quarks (up, down, strange, charmed, bottom and top), and each flavour comes in three 'colours' (red, green or blue). These terms are, of course, simple labels assigned by physicists to distinguish the various types of quark; for example, being vastly smaller than the wavelength of visible light, they do not have colour in the normal sense of the word.

It is now known that protons and neutrons are each composed of three quarks, one of each colour. A proton consists of two up quarks, each with a charge of $+\frac{2}{3}$ (of the electronic charge) and one down quark with a charge of $-\frac{1}{3}$, giving it a net charge of $+1$. A neutron contains two down quarks and one up quark and is therefore electically neutral. Other combinations of quarks with various flavours and colours form different particles, but these have short lifetimes and decay rapidly into protons and neutrons. Up, down and strange quarks were discovered in the 1960s, the charmed quark in 1974 and the bottom quark in 1977. The top quark was found only recently – in 1995.

Is it conceivable that quarks themselves are further divisible into even smaller particles? Recent experiments suggest that they may have some structure, or interact in a hitherto unsuspected way, although there are theoretical reasons for believing that quarks and electrons represent the basic building blocks from which all matter is made. What is clear is that machines capable of accelerating particles to higher energies than have currently been achieved will be required to isolate smaller particles, if indeed they exist.

The various types of quarks, together with electrons, form a gallery of what might be called *matter particles*. There is, however, another category of particles used to account for the interactions or forces between matter particles. The modern view is to describe everything in the universe, including light, gravity and the forces between elementary particles, in terms of particles. This follows from the concept of wave/particle duality which quantum mechanics and quantum field theory tell us applies universally. Thus, for example, light can be considered as an electromagnetic wave or a stream of particles named *photons*.

The assumption that light (using the term here in a general sense to describe electromagnetic radiation both within and outside the visible spectrum) is emitted from a source in tiny packets of energy was first put forward by Planck in 1900 but without any real conception of why it should be thus quantised. In 1905 Einstein postulated that quantisation was a fundamental intrinsic property of light, not connected with the emission process itself. His theory, proposed to explain the photoelectric effect in which electrons are ejected from a metal by light incident upon its surface, initially received a hostile reception, mainly because of its similarity to Newton's corpuscular theory of light which had been decisively rejected in the nineteenth century in favour of the rival wave theory. Further evidence for the photon[2] as a particle aspect of electromagnetic radiation came in 1923, when Compton found that light scattered from protons underwent a change of wavelength corresponding to the photons losing energy in the collision process.

The photon can also act as a *force-carrying particle*, responsible for the interaction (attractive or repulsive) between elementary matter particles. In this role it is called a 'virtual' photon. For example, the electric repulsive force between two electrons can be considered as arising from an exchange of virtual photons between the two charges. There are other force-carrying particles. Unlike matter particles which have a spin[3] of $\frac{1}{2}$, force-carrying particles have spins of 0, 1 or 2. Photons have a spin of 1 and, as we have seen, are the particles associated with electromagnetic fields or forces. The forces which hold protons and neutrons together in a nucleus are not of this type; neither are those which bind quarks together in a proton or a neutron. Nuclear forces have a short range but must be more powerful than that associated with, for example, electrical repulsion between two protons, otherwise nuclei would fly apart. These so-called *strong forces*, which operate on nuclear scale lengths, are carried by another particle of spin 1 called (appropriately) a *gluon*, which, like quarks, have colour. Individual gluons and quarks do not exist in isolation because it is known that naturally occurring particles must be colourless. By analogy with the combination of the three primary colours producing 'white', protons and neutrons must be composed of equal numbers of red, blue and green quarks (in total, three quarks, not two or four) held together by gluons.

In addition to the photon and the gluon, there are other force-carrying particles, which were predicted theoretically before being discovered experimentally. They are the W^+, W^- and Z^o particles and are the 'mediators' or 'quanta' associated with the so-called *weak* nuclear force which is responsible for radioactive beta decay. These particles are approximately 100 times as massive as the proton or neutron but are very short-lived.

Studies of beta decay processes led first to the postulate, and then the discovery, of another particle called the *neutrino*, carrying no charge and a spin of $\frac{1}{2}$ but having zero mass. In beta decay, neutrons are converted to protons with the emission of an electron (the β-rays of radioactivity), transforming one element into another; for example, potassium ($^{40}_{19}K$) into calcium ($^{40}_{20}Ca$). (The upper number is the number of neutrons and protons in the nucleus; the lower, the number of protons.) The emitted electrons vary considerably in energy, which led Pauli to suggest the emission of a 'silent accomplice' to satisfy the law of conservation of energy. In 1933 Fermi presented a theory of β-decay and gave the name neutrino to the elusive particle. In the 1940s, the neutrino was found to play a role in a decay process involving two other fundamental particles, the *pion* and the *muon*, discovered in studies of cosmic rays at high altitudes. The *pion* has a mass intermediate between the electron and the proton and for this reason is called a *meson*, meaning middle-weight. In the same spirit, an electron is called a *lepton* (light-weight) and protons and neutrons are *baryons* (heavy-weight). The pion decays to a muon with the emission of a neutrino and the muon itself is unstable. Whereas the pion is now known to be composed of quarks, the muon is an elementary particle in its own right and, although over 200 times heavier than an electron, belongs to the lepton family, which also includes the neutrino. The meson family is much larger, the pion being only one of many particles, each of which is composed of various combinations of quarks and antiquarks.

The concept of *antiparticles* was introduced by Dirac in 1928, whose theory of the electron and other particles carrying a spin of one-half, predicted the existence of a largely unseen world in which matter consists of antiparticles, i.e. particles having the same mass as the normally observed particle but with opposite charge. Dirac's prediction of positively charged 'electrons' was vindicated in 1931 with the discovery of the *positron* – the twin of the electron. Subsequently other antiparticles were found in high-energy collision processes. Particles and antiparticles mutually annihilate when they meet, the energy appearing as radiation. Cosmology attempts to explain why, if particles and antiparticles were created in equal numbers at an early stage in the creation of our universe, an imbalance developed later which permitted the condensation of primordial gas into galaxies, stars and planets, composed only of one type of these particles.

Allowing for the facts that force-carrying particles are their own antiparticles, and that several hundred known mesons and baryons can be decomposed into quarks, we are still left with a very large number of *elementary* particles: the present count is 12 leptons, 12 types of quark (each of which can have one of three colours) and 12 mediators, with at least one more (the Higgs particle) being called for by recent theories. All this seems a long cry from the discovery of the first lepton by Thomson 100 years ago.

The Electronics Revolution

The story of elementary particles as told in the above outline is unknown to most non-physicists. In contrast, the practical realisation and utilisation of electronics has had an impact on everyday life which is familiar to everyone. It parallels that of the industrial revolution of earlier centuries.

Usage of electricity in industry predates the discovery of the electron by about 50 years and indeed at the time Thomson was undertaking his revolutionary experiments in the 1890s, the major cities of the world were lit by electricity, transatlantic telegraph cables had been laid, electric trams were running, and electrical machinery in the factory and appliances in the home were becoming commonplace. One did not need to know the *nature* of electric currents to use them effectively.

The first attempts to create electronic devices based on the motion of electrons and control of their flow were stimulated by the discovery of the transmission of long-wavelength electromagnetic radiation through space by Hertz in 1888 and the promise that this offered for radio signalling. In 1896 Marconi, then only 20 years of age, came to England from Italy and, backed by Sir William Preece of the Post Office, developed transmitters and receivers which permitted wireless communication over greater and greater distances. By 1901, signalling over several hundred miles had been achieved and, a few years later, use of longer wavelengths and higher powers increased this distance to a few thousand miles. By 1907, radio-telegraphy was firmly established on a commercial basis.

The first valves were designed and patented by Fleming in 1904 as detectors of radio waves. Known as thermionic diodes, they served to rectify the oscillatory signals received by the aerial and frequency-selected by the tuning circuits of early radio receivers. They replaced coherers, crystal detectors and other devices that had been used previously for this purpose. Fleming's valves were essentially light bulbs (invented by Edison in 1883) in which the incandescent filament was surrounded by a cylinder of sheet or gauze. In these devices, when the cylinder is at a positive voltage with respect to the filament, electrons emitted from the filament are readily collected by the cylindrical anode and a current flows, but with the opposite polarity they are turned back by the field and the current ceases. Thus oscillatory signals fed to the device are converted into voltage excursions of one polarity only and the rectified signal is then suitable for detection by, say, earphones.

In 1907 de Forest designed the triode valve in which a third electrode in the form of a cylinder wire helix, known as the grid, was positioned between the filament and the anode. A current flowing through a triode valve can be controlled by a voltage applied to the grid and, most usefully, small oscillatory signals applied between the filament and the grid can be amplified into larger excursions of voltage at the anode. Variations in the design of triodes (for example, the insertion of extra grids in the so-called tetrode or pentode valves) was found to improve the stability and extend the frequency response for amplification. Valves were used as components of electrical circuits not only as amplifiers but as switching elements, oscillators and for impedance-matching purposes until the 1950s when they were largely replaced by solid-state devices. The first electronic computers contained thousands of valves: for example, Colossus, the first electronic computer, built at Bletchley Park during the Second World War for the purposes of breaking the codes of the German Enigma machine, had 1500 valves; and ENIAC, the Los Alamos National Laboratory computer used to simulate H-bomb explosions, contained 18 000.

For the generation of ultra-high frequencies, valves in which the electrons moved in a magnetic as well as an electric field were developed, again during the war years for radar installations. One such device, the cavity magnetron – invented by Randall and Boot – is capable of very high power, high-frequency, pulsed output.

The cathode ray oscilloscope, used for the visual display and analysis of electrical waveforms, has a vacuum tube which is a direct descendant of that used by Thomson for the determination of e/m for the electron and has essentially the same form. Electric and magnetic fields are used to steer a beam of electrons, emitted from a cathode and passing through a cylindrical anode, prior to their impingement on a fluorescent screen. Television tubes and computer monitors provide examples of the development of the cathode ray tube from its early utilisation as a scientific instrument to a more generally used visual display unit.

Electron optics, involving precision focusing of electron beams by either magnetic or electrostatic lenses, found important application in the electron microscope, the first of which was constructed in the 1930s. An optical microscope cannot resolve details of an object having linear dimensions less than the wavelength of visible light, thus limiting its useful magnification to about 1000. In the mid-1920s, J.J. Thomson's son, G.P. Thomson, demonstrated that fast electrons could be diffracted in much the same way as X-rays or other electromagnetic radiation – a manifestation of the wave-particle duality principle of quantum mechanics. The short wavelength of electrons accelerated through a potential difference of, say, 50 000 volts is used in electron microscopes to achieve magnifications of several hundred thousand. Resolution of individual atoms on the surfaces of suitably prepared samples is now quite commonly achieved in the transmission electron microscope (TEM). In the scanning electron microscope (SEM), the object is scanned with a raster of primary electrons and emitted secondary electrons are used to form the image. The first SEM was built in 1951 and commercial instruments became available in 1965. A descendant of the SEM – the scanning probe microscope (SPM) – has a magnification of about one million and permits not only imaging but also the manipulation of individual atoms for the purposes of microfabrication of tiny electronic devices.

Another instument utilising the principle of electrostatic deflection is the high-speed electronic camera. Light from the event to be photographed is imaged optically onto a photocathode which emits electrons. These pass through sets of parallel plate electrodes which carry rapidly varying voltages for the purpose of scanning the image across a diaphragm in front of a conventional photographic film. Modern versions (e.g. the *Imacon*) are capable of recording 20 million frames per second. Steering of electrons and amplification by secondary electron emission is used in photomultiplier tubes for the detection of low levels of light.

Discharges through gases, a topic which occupied so much of Thomson's time, have found applications in lighting, with different gases being used according to the requirements. Neon, sodium and mercury are the most common, but lamps having virtually any chosen spectral output can be obtained using gaseous mixtures. The electrodeless discharge, first demonstrated by Thomson, has recently been developed by the General Electric Company and a lamp – the *GE Genura* – marketed for domestic purposes.

Rectifiers, photocells and other solid-state devices, made from selenium or copper oxide, were used in electronic applications prior to a full understanding of the way they operated. It was, however, the discovery of the transistor in 1947, by Bardeen, Shockley and Brattain at the Bell Telephone Laboratories in the USA, which ushered in a new era of electronics based on semiconductor materials. As the name suggests, semiconductors are substances having electrical conductivities lying between those of an insulator and a metal. It is not, however, this property that is important but rather the ability to *change* the conductivity by the addition of small quantities of other elements to an otherwise pure semiconductor – a process known as *doping*. Doping can be of two kinds: the addition of a *donor* element creates *n-type* material, in which the dominant carriers of electricity are *electrons*; whereas the addition of an *acceptor* element produces *p-type* material in which the predominant carriers are *missing electrons* or *holes*. In both pure and doped semiconductors, the electrical conductivity depends on the temperature, and this property is utilised in some device applications; in others it is the sensitivity of the electron and/or hole concentrations to incident radiation (say light) or to magnetic fields which provides useful characteristics. In all cases, the starting material needs to be purified to a degree such that no more than say 1 part in 10^{10} of impurities is present. Germanium was the first element to be utilised in devices but this has now been largely replaced by silicon and, in some cases, by compound semiconductors such as gallium arsenide.

The first solid-state devices fabricated from semiconductors attempted to mimic the operations of existing vacuum tube devices. For example, a junction formed between n-type and p-type material simulated the behaviour of the thermionic diode and an npn or pnp sandwich could be made to operate like a triode. The obvious advantages of solid-state devices are their small size, their cheapness and reliability, and their low power consumption, all of which permitted what was at the time an unprecedented degree of miniaturisation and high packing density. These qualities have been further developed in the *integrated circuit*, in which discrete components, such as diodes, transistors, resistors and capacitors, along with their interconnections, are fabricated by lithographic and various chemical processes on a single wafer of, say, silicon. Today, such 'chips' form the heart of all circuits in computers, radios, televisions, electrical instruments and consumer products.

Semiconductors are also used in light-detecting devices and in photovoltaic applications such as solar cells which convert light into electricity. The converse process – conversion of electricity into light – is used in light-emitting diodes (LEDs) and solid-state lasers. Other applications are in image convertors, photocopiers, fax machines and other scanning devices. Flat display panels of the active matrix type use millions of transistors to switch light on or off at individual pixels, and are replacing the cathode ray tube for visual displays of all kinds.

As the present century draws to a close, there is no sign of any slackening in the rate of development or design of new electronic devices and in their miniaturisation. Digital switching in microprocessors, for example, can now be accomplished by the motion of *single* electrons. The discovery of new applications will also no doubt continue, affecting almost all aspects of human activity and exerting a profound impact on society in general.

Notes

1 The origin of the name is an enigmatic quotation from *Finnegan's Wake* by James Joyce: 'Three quarks for Muster Mark!'

2 The name *photon* was suggested by the chemist Gilbert Lewis in 1926.

3 A classical way of picturing spin is to imagine a particle as a sphere rotating about an axis. An alternative viewpoint is to consider what the particle looks like from different directions.

Index

The page numbers in square brackets refer to the facsimiles. Keywords in the facsimiles are not individually indexed.